WALCH PUBLISHING

Daily Warm-Ups

MATH
BRAIN TEASERS

Susan Conover

Level I

The classroom teacher may reproduce materials in this book for classroom use only.
The reproduction of any part for an entire school or school system is strictly prohibited.
No part of this publication may be transmitted, stored, or recorded in any form
without written permission from the publisher.

1   2   3   4   5   6   7   8   9   10
ISBN 0-8251-5504-5
Copyright © 2005
J. Weston Walch, Publisher
P.O. Box 658 • Portland, Maine 04104-0658
walch.com
Printed in the United States of America

Table of Contents

Introduction   iv

The **Daily Warm-Ups series** is a wonderful way to turn extra classroom minutes into valuable learning time. The 180 quick activities—one for each day of the school year—review, practice, and teach math concepts. These daily activities may be used at the very beginning of class to get students into learning mode, near the end of class to make good educational use of that transitional time, in the middle of class to shift gears between lessons—or whenever else you have minutes that now go unused.

Daily Warm-Ups are easy-to-use reproducibles—simply photocopy the day's activity and distribute it. Or make a transparency of the activity and project it on the board. You may want to use the activities for extra-credit points or as a check on the math skills that are built and acquired over time.

However you choose to use them, *Daily Warm-Ups* are a convenient and useful supplement to your regular lesson plans. Make every minute of your class time count!

Numbers, Numeration, Operations, and Patterns

Name a two-digit number less than 1 (that is, to the right of the decimal point) whose mirror image is a correctly written whole number.

1

© 2005 Walch Publishing

Numbers, Numeration, Operations, and Patterns

While washing up for dinner, Bernie notices something very peculiar about the bar of soap and the washcloth. With one swipe of the soap on the washcloth, he creates 10 bubbles. With another swipe, he creates 100 bubbles, yet he pops half of the first swipe's bubbles. On the third swipe, he makes 1,000 bubbles, and half of the second swipe's bubbles are popped.

If this pattern continues, how many bubbles will there be after the sixth swipe of the bar of soap?

2

What is the number that is 7 more than one third of one fifth of one tenth of 3,750?

Daily Warm-Ups: Math Brain Teasers

Numbers, Numeration, Operations, and Patterns

Assume that the number 2,100 is the result of a number being rounded. What numbers could have been the original number?

4

Numbers, Numeration, Operations, and Patterns

Anya was offered a temporary job. The employer said, "We need you for 7 days. I can give you a choice of payment plans. You can get a flat rate of $20 a day. Or you can start at $2.00 a day and then have your salary double every day for the course of the week."

Which payment plan should Anya choose?

5

Numbers, Numeration, Operations, and Patterns

The cat king of Purrdom told his loyal felines, "I have the power to grant you 1 more life in addition to your current 9 lives, but there is a catch. You must put the digits that make up your 9 lives together in such a way that they make 27."

How might the cats make 27?

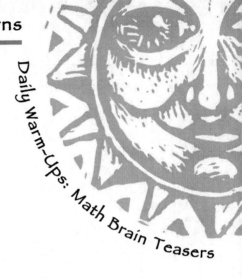

6

Numbers, Numeration, Operations, and Patterns

A medieval mathematician and businessman by the name of Leonardo Pisano (1170–1250), also known as Fibonacci, came up with the following sequence of numbers to describe phenomena in nature, art, geometry, architecture, and music.

What is the next number in the Fibonacci sequence?

1, 1, 2, 3, 5, 8, 13, 21, 34, 55, . . .

7

Numbers, Numeration, Operations, and Patterns

The Mayans of Mexico used a method of dots and lines to show a period of time. Each dot represents a unit, and each bar represents 5 units. From the bottom position to the top position, the symbols refer to the number of days, months, and years, respectively. The example below shows 15 years, 11 months, and 8 days.

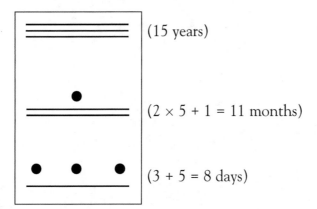

(15 years)

$(2 \times 5 + 1 = 11$ months$)$

$(3 + 5 = 8$ days$)$

If today is April 23, 2006, write the exact age of a person born on March 9, 2001, using the Mayan method of calendar time. (Assume there are no leap years.)

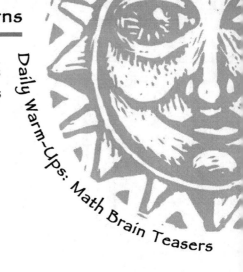

8

Numbers, Numeration, Operations, and Patterns

The Chinese calendar is based on a 12-year cycle. Each year in the cycle is associated with an animal: rat, ox, tiger, rabbit, dragon, snake, horse, sheep, monkey, rooster, dog, and pig. The order of the animals never changes.

If the rooster marks the year 2005, what animal will mark the year 2068? _____

9

Numbers, Numeration, Operations, and Patterns

A farmer has a super-chicken named Harriet. Harriet can lay 10 eggs at one time, all of which hatch female chicks. When the chicks grow up, they too will be able to lay 10 eggs that hatch hen chicks, and so will the generations to follow.

The farmer numbers his chicks consecutively. Assuming that each hen takes a break after laying her first 10 eggs, how is chick number 2 related to chick number 602?

10

Numbers, Numeration, Operations, and Patterns

Prime factorization is finding the factors of a number that are all prime numbers.

To find the prime factors of a number, you start by finding two factors. If either factor is not a prime number, find factors for that factor. Continue the process until all your factors are prime.

Find factors and multiples that will correctly complete the prime factorization diamond below. *Hint:* The same number appears in the boxes at both ends.

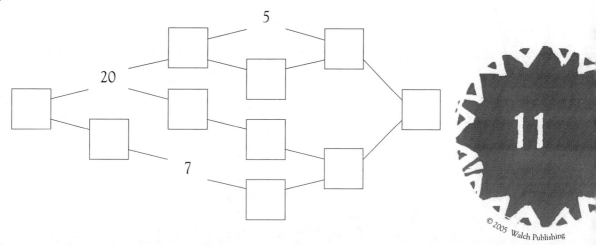

Numbers, Numeration, Operations, and Patterns

The Wizard Wesley told his young apprentice that the gates to his castle can only be opened by creating a palindrome (a word or a number that reads the same forward and backward—for example, Hannah or 545). The wizard said, "Using the digits 1 through 4, find two 4-digit numbers that are palindromes and whose sum is a palindrome."

The young apprentice successfully opened the gates to the castle. Name two numbers that she might have used.

_____ and _____

12

Numbers, Numeration, Operations, and Patterns

Carmen is planning a neighborhood block party for 50 guests. She needs to buy hot dogs and hot dog buns. Inconveniently, hot dogs come in packages of 10, and buns come in packages of 8.

How many packages of hot dogs and hot dog buns should she buy so that each of her guests gets the same number of hot dogs on a bun? (In other words, assume no leftovers.)

13

Numbers, Numeration, Operations, and Patterns

Melinda, Melony, Melody, and Melana are playing a game much like the television game show *Jeopardy*. The scorekeeper, Melicent, spilled her coffee on the scorecard. The names on the scorecard were blotted out, but the scores are still visible. Melicent remembers the following about the scores: Melody had a low score. Melana's score was greater than Melony's score. Melony's score was 100 points greater than Melinda's score.

Which score belonged to which girl?

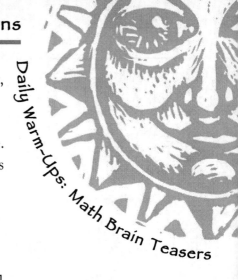

14

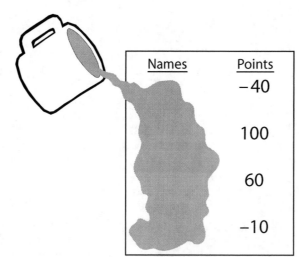

Names	Points
	−40
	100
	60
	−10

Numbers, Numeration, Operations, and Patterns

Harvey forgot to bring his glasses to school. As a result, he did not see the decimal points in an equation in his notebook. He copied the equation onto the board as follows:

2585 + 190 + 4631 = 7235

Help Harvey by finding where the decimal point should go in each of the addends and the sum to make the equation true. Write the correct equation below.

_____

15

© 2005 Walch Publishing

Numbers, Numeration, Operations, and Patterns

Mallard is a very orderly dog. In order, he chews on a bone, a shoe, a piece of bark, and a rubber steak. He then repeats this sequence. If he repeats the whole sequence several times, what will be the twenty-second item chewed on?

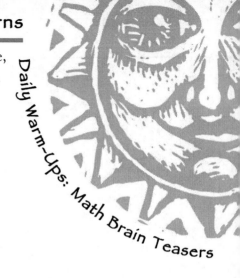

16

Numbers, Numeration, Operations, and Patterns

Find the missing digits in the multiplication problem below.

$$
\begin{array}{r}
8\square2 \\
\times\ \square64 \\
\hline
33\square8 \\
4\square920 \\
\square32800 \\
\hline
\square86{,}0\square8
\end{array}
$$

© 2005 Walch Publishing

17

Numbers, Numeration, Operations, and Patterns

Nester, a UFO fanatic, had a dream in which he was abducted by 11 little purple aliens. There were two types of purple alien: Byopics, which have 2 pink eyes, and Tryopics, which have 3 pink eyes.

If he remembers seeing 23 pink eyes, what is the greatest number of Byopics Nester could have dreamed he saw?

18

Daily Warm-Ups: Math Brain Teasers

Carl Friedrich Gauss (1777–1855) was a famous mathematician. When Gauss was young, his teacher came up with a project to keep the class busy. He told them to find the sum of all the numbers between 1 and 100. Within a minute, Gauss handed in the correct answer. Instead of adding all the numbers one after the other (1 + 2 + 3 + . . .100), he saw a pattern. The numbers could be grouped into pairs, each of which added up to 101 (1 + 100, 2 + 99, 3 + 98 . . .). There were 50 of these pairs. Gauss multiplied 101 × 50, for a total of 5050.

Use Gauss's method to find the sum of the numbers 1 to 199.

19

Numbers, Numeration, Operations, and Patterns

6 is considered a perfect number because its factors (excluding 6 itself) add to 6.

There is a perfect number between 20 and 30. What is it?

_____

20

Numbers, Numeration, Operations, and Patterns

An ancient Hindu mathematician named Aryabhata liked to do math puzzles. His method of solving puzzles is called *inversion*. This means to work backward, or do the opposite operations. Use inversion to solve the following just as Aryabhata would have done.

A number is divided by 2.5. The answer is multiplied by 5 and is added to $16\frac{9}{10}$, and the result is 18. What is the original number?

21

Numbers, Numeration, Operations, and Patterns

Silas is 4 years old. His sister Opal is 3 times as old as he is.

When Silas is 12, how old will Opal be?

22

Numbers, Numeration, Operations, and Patterns

The decimal points have been left out of both the divisor and the dividend in this problem.

28035 ÷ 45 = 62.3

Without adding zeros to the divisor, what are all the possible divisors and dividends that make the quotient true?

23

Numbers, Numeration, Operations, and Patterns

A boy was standing on the shore of a lake. He threw a rock into the water. A grumpy old fish, not liking to be disturbed in this manner, threw 2 rocks back onto the beach. The boy threw another rock in the lake. This time the old fish threw 3 rocks back onto the beach. This continued with increasing fury from both the boy and the fish.

How many turns (one turn being that the boy throws a rock and then the fish throws one more rock than it did the last time) must take place until there are more than 50 additional rocks on the beach?

24

Numbers, Numeration, Operations, and Patterns

About how long, in years, would it take to count to a billion, if each count took 1 second?

25

Numbers, Numeration, Operations, and Patterns

With a little imagination, you can write math terms so that they show what they mean. For example, exponents show the power to which a number is raised. So we might write the word *exponents* as

e^{x}p^{o}n^{e}n^{t}s.

Try your hand at being clever. Think of a fun way to write the following words.

decimals   rounding

integers   multiplying

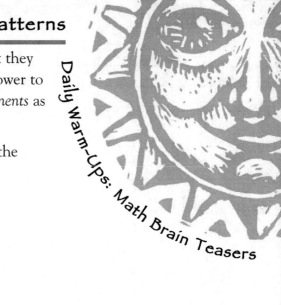

Daily Warm-Ups: Math Brain Teasers

26

Numbers, Numeration, Operations, and Patterns

A math term is expressed by the drawing below. Can you figure out what it is?

27

Numbers, Numeration, Operations, and Patterns

Look at each row of numbers in the pyramid. What number should replace the question mark in the middle of the bottom row?

```
                8
             2  5  2
          1  2  4  2  1
       1  2  1  3  1  2  1
    1  2  1  1  ?  1  1  2  1
```

28

Numbers, Numeration, Operations, and Patterns

A circle of 11 trees is planted, with the trees close together. Over time, their roots become entangled with one another. Each tree has one—and only one—root entangled with one root from every other tree.

How many tangles are there?

29

Numbers, Numeration, Operations, and Patterns

A perfect square is a number whose square root is a whole number. For example, 9 is a perfect square, because the square root of 9 is 3.

Find two perfect squares larger than 100 but less than or equal to 300, whose sum is a perfect square.

30

If the digits –5, 5, –4, 4, –3, and 3 are arranged on the points of the triangle so that the sums of the digits on each triangle are opposite numbers, how many possible combinations of digits are there, if no combination is repeated?

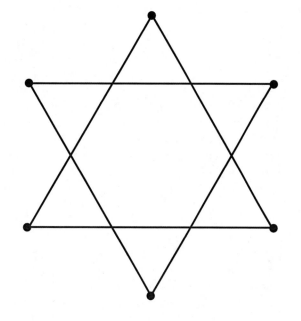

© 2005 Walch Publishing

31

Numbers, Numeration, Operations, and Patterns

To what power must 2 and 3 be raised in order to equal each other?

32

Daily Warm-Ups: Math Brain Teasers

Numbers, Numeration, Operations, and Patterns

Using only the digits 0, 1, 3, 4, and 9, make as many equations using addition as you can. The digits must be used for both the numbers being added and the sum, and you may use each digit only once in each equation.

33

© 2005 Walch Publishing

Numbers, Numeration, Operations, and Patterns

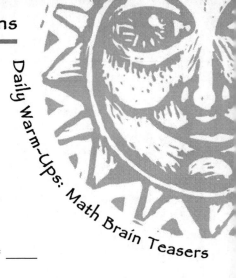

For this exercise, the numbers 1 to 26 correspond to the letters A to Z. First, find the missing numbers by solving the math problems below. Then write the letter that corresponds to each number to find the hidden message.

$$\sqrt{169} = \underline{\qquad} \quad 15 + -14 = \underline{\qquad} \quad 4 \times 5 = \underline{\qquad} \quad -2 \times -4 = \underline{\qquad}$$

$$3^2 = \underline{\qquad} \quad 33 - 14 = \underline{\qquad}$$

$$-75 \div -25 = \underline{\qquad} \quad 5.6 + 9.4 = \underline{\qquad} \quad 33 \div 2.2 = \underline{\qquad} \quad 16.6 - 4.6 = \underline{\qquad}$$

Numbers, Numeration, Operations, and Patterns

Look at the pattern below.

What will the twelfth square look like? Draw it in the space below.

   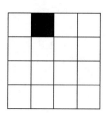 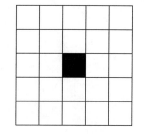

Geometry and the Coordinate Plane

Draw a star with seven points without lifting your pencil or retracing lines.

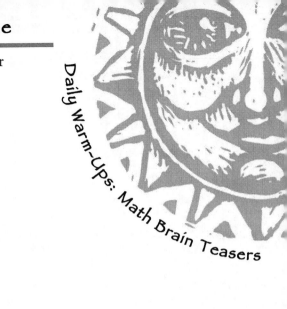

36

Geometry and the Coordinate Plane

How many of the letters in the Greek alphabet have more than one line of symmetry?

$$A \; B \; \Gamma \; \Delta \; E \; Z \; H \; \Theta \; I \; K \; \lambda \; M \; N \; \Xi \; O \; \Pi \; P \; \Sigma \; T \; Y \; \Phi \; X \; \Psi \; \Omega$$

37

Geometry and the Coordinate Plane

A caterer is setting up for a party and is trying to arrange the tables. He needs a total of 10 tables. The party-giver wants the tables arranged in rows of 4.

How can the caterer set up 10 tables so that each one is in a row of 4?

38

Geometry and the Coordinate Plane

It is said that a picture is worth a thousand words. Sometimes, you can make words act like pictures. This shows their meanings visually. For example, perpendicular lines are lines at right angles to each other. So you might write the term like this:

perpendicular lines

Try your hand at being clever. Think of a fun way to write the following words.

obtuse

acute

equilateral triangle

perimeter

volume

39

Geometry and the Coordinate Plane

A group of people are standing in a circle. They are spaced evenly apart. The ninth person is directly opposite the twenty-fifth person.

How many people are in the circle?

40

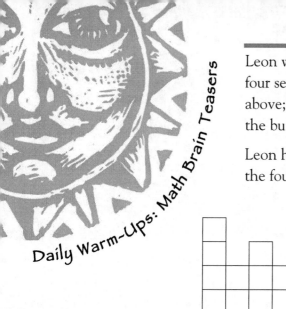

Geometry and the Coordinate Plane

Daily Warm-Ups: Math Brain Teasers

Leon was preparing drawings for a new building. He had to draw four separate views: one from directly in front; one from directly above; one directly from the side; and one at an angle, showing the building in perspective.

Leon has done the first three views. Can you combine them to do the fourth view, showing the completed building?

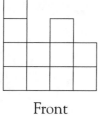

Front

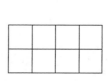

Top

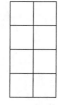

Side

41

© 2005 Walch Publishing

Geometry and the Coordinate Plane

Wilma is a designer for an outdoor goods manufacturer. She has been asked to create a pattern for a new tent. Here is the tent design. Create a net that, when folded, will form this shape.

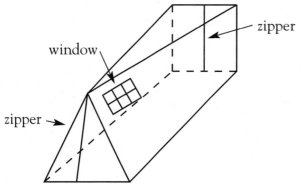

window

zipper

zipper

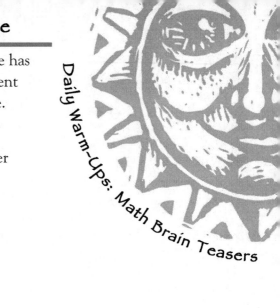

42

Geometry and the Coordinate Plane

Look at the circle below. Draw three straight lines across the circle to divide it into sections, with only one small square in each section.

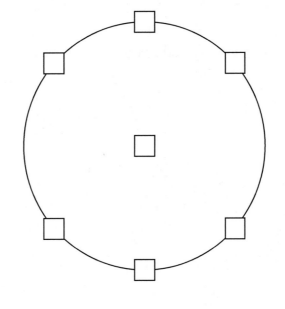

Daily Warm-Ups: Math Brain Teasers

43

Geometry and the Coordinate Plane

An *analogy* describes a relationship between items. For example,

black : white :: day : night

(Black is to white as day is to night.)

In this case, the first part of the analogy describes a relationship of opposites. The second part must then describe a comparable relationship.

Now try some on your own.

1. triangle : equilateral :: rectangle : _____

2. snowman : 8 :: an acute angle : _____

44

Geometry and the Coordinate Plane

What should be the next symbol in the series below?

II Ω2 Ɛ3 ᚻ4 ᘓ5 ᘓ6

Geometry and the Coordinate Plane

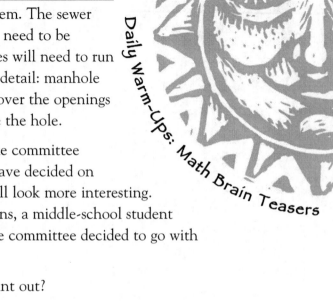

A small town is setting up its first-ever sewer system. The sewer committee has worked out how many houses will need to be hooked up to the system, and what route the pipes will need to run underground. Now they are working on the final detail: manhole covers. These are the metal plates that form lids over the openings of manholes. The cover rests on a small lip inside the hole.

Traditionally, manhole covers are round. But some committee members want to do something different. They have decided on rectangular manhole covers, which they think will look more interesting. However, at a public meeting to display the designs, a middle-school student pointed out a problem in their design. The committee decided to go with round covers after all.

What problem did the student point out?

46

Geometry and the Coordinate Plane

A new reservoir has recently been built, covering hundreds of acres with water. Near its shore, the top of a flagpole still sticks out of the water. One tenth of the pole is buried in the ground under the lake. One half of the pole is covered by water. At the top, 10 feet of the pole sticks out of the lake.

What is the total height of the pole, in feet?

47

Geometry and the Coordinate Plane

Pagoda is an ancient Asian game. The object of the game is to move discs from one peg to another so that the discs are arranged from largest to smallest. The rules are that only one disc can be removed at a time, and a large disc may not be placed on top of a smaller disc.

What is the smallest number of moves it will take to move the discs to the peg on the right?

48

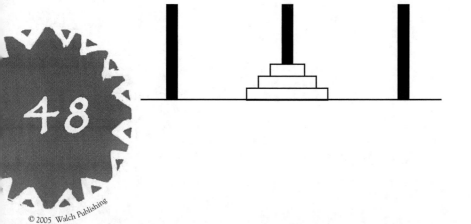

Geometry and the Coordinate Plane

Two spiders are making their way to two flies at the same time. Spider A is tracking down fly A, and spider B is tracking down fly B. Which spider will get to its fly first?

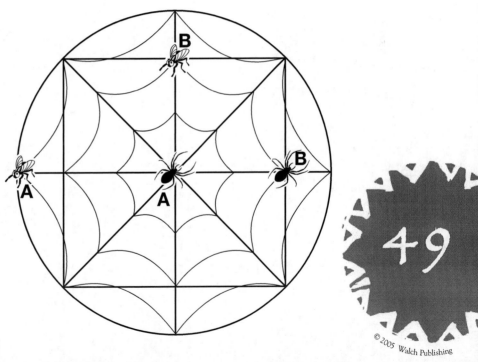

49

Geometry and the Coordinate Plane

Zipper, a cat with no common sense, is stuck 12 feet up a telephone pole. If a ladder is placed at a 45° angle against the telephone pole, what must be the length of the ladder, in feet, to reach Zipper?

50

Geometry and the Coordinate Plane

What is the area of the puzzle if each square puzzle piece is 5 centimeters long?

51

Geometry and the Coordinate Plane

Find the area of the figure, in square units.

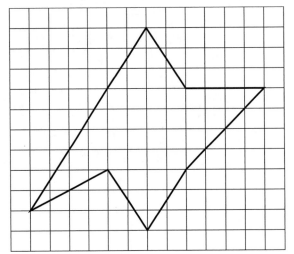

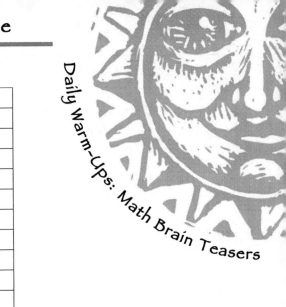

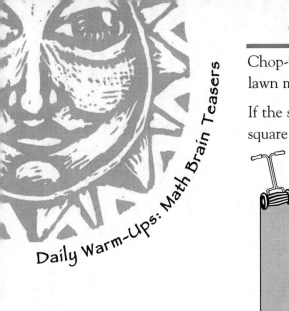

Geometry and the Coordinate Plane

Chop-Chop Lawn Care Service is mowing a square lawn. The lawn mower has a 4-foot-wide cutting deck.

If the service needs to go around the lawn 9 times, what is the square footage of the lawn?

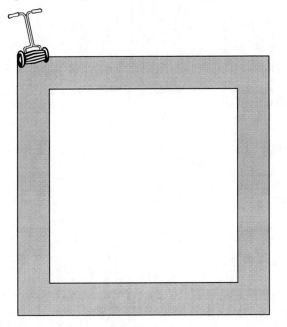

Divide the figure into two congruent shapes.

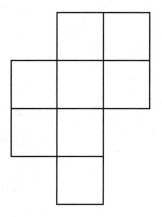

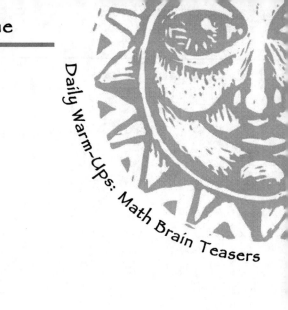

Geometry and the Coordinate Plane

Imagine that the dotted line below is a mirror. On the other side of the line, draw a mirror image of the drawing.

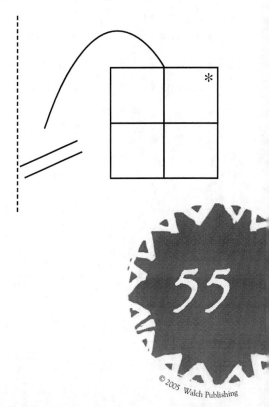

Geometry and the Coordinate Plane

An astronomer was plotting the coordinates of a space satellite when it was hit by an asteroid. This caused the satellite to drift clockwise on the coordinate plane to quadrant III, stopping at the same distance from the origin as it was in quadrant I.

Plot the coordinates for the satellite's new position. Then draw the satellite where it should be in quadrant III.

56

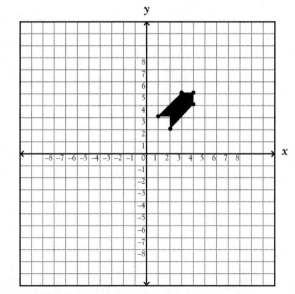

Find the similar figures.

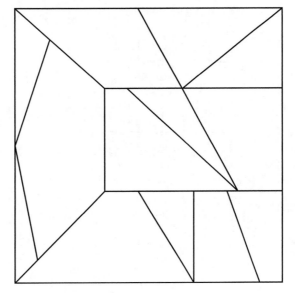

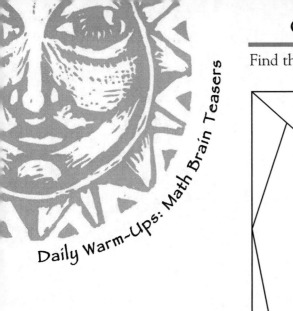

57

Geometry and the Coordinate Plane

What is the perimeter of this figure, in units? (*Hint:* Use 3.14 as a value for π.)

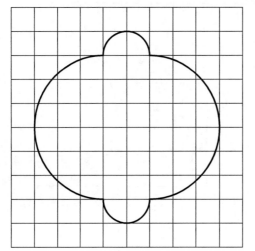

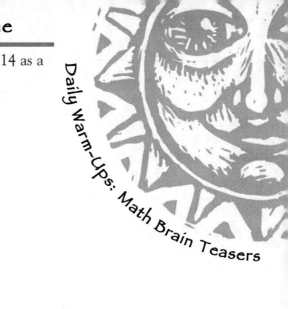

58

Find the measure of the angle. (*Hint*: The triangle is equilateral.)

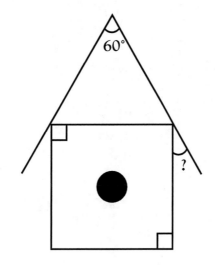

59

Geometry and the Coordinate Plane

What is the measure of the exterior angle of the isosceles-shaped tepee?

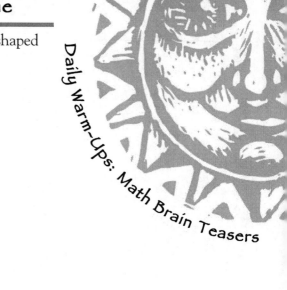

25°

?

60

© 2005 Walch Publishing

Daily Warm-Ups: Math Brain Teasers

Geometry and the Coordinate Plane

The drawing below should suggest a geometry term. What is it?

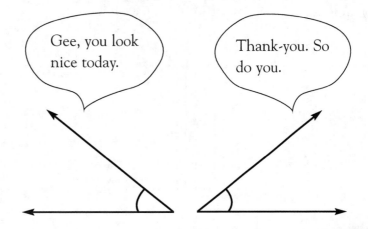

Gee, you look nice today.

Thank-you. So do you.

61

Geometry and the Coordinate Plane

The note card below is attached to a string. If the card is spun rapidly around the string, what would it look like?

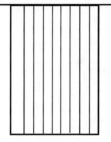

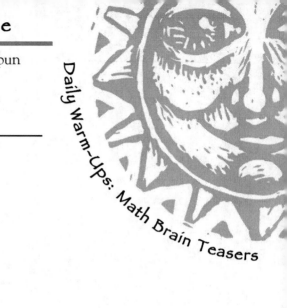

62

Geometry and the Coordinate Plane

Area and perimeter are two different things. But once in a while, they are similar. A square of certain dimensions has a perimeter that is the same in linear units as its area is in square units.

What are the dimensions of this square?

Geometry and the Coordinate Plane

For which figure is the area greater than the perimeter?

a.
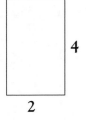
4

2

b.
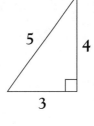
5
4

3

c.
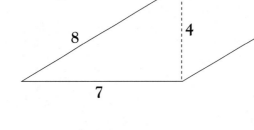
7

8

4

8

7

d.
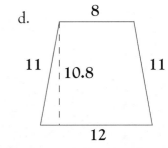
8

11

10.8

11

12

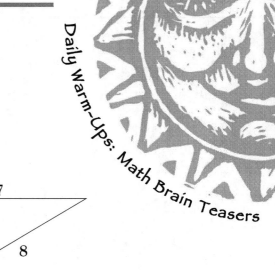

64

© 2005 Walch Publishing

Geometry and the Coordinate Plane

The three pickle slices below were all made by slicing the same pickle. Draw lines on the pickle to show where it should be cut to give each shape.

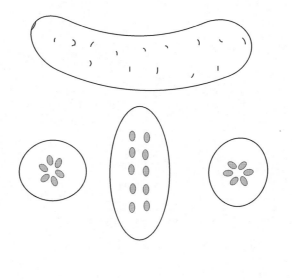

Geometry and the Coordinate Plane

Arif needs to fill his ice chest with 2-inch ice cubes. What is the greatest number of ice cubes that will fit?

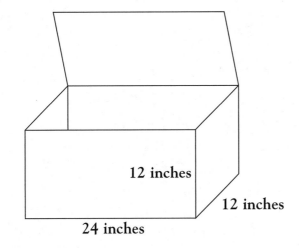

12 inches

12 inches

24 inches

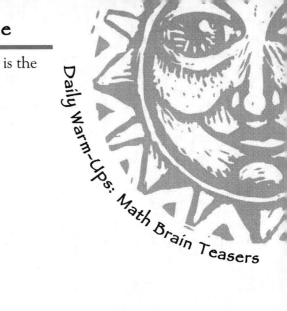

66

What is the area of the iris?

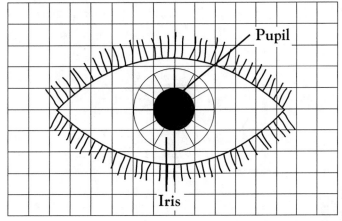

Pupil

Iris

67

Daily Warm-Ups: Math Brain Teasers

Geometry and the Coordinate Plane

A *fractal* is a geometric pattern that is repeated at increasingly smaller or larger scales. The first three drawings below show three stages of a fractal based on triangles. Look at them carefully. Then draw the next two stages for the fractal at the bottom of the page.

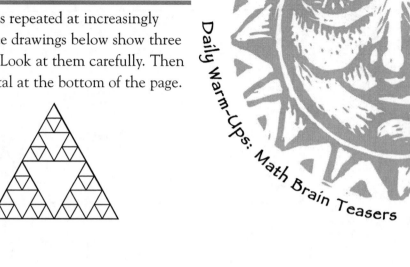

Geometry and the Coordinate Plane

Juride is a game that uses dots. The objective of the game is to draw a closed loop that includes all the dots. The rules are that you may not retrace a line segment, no diagonals may be made, and there must be at least one opening on each end of the four sides of the square.

Try playing juride with this 6-by-6 square of dots.

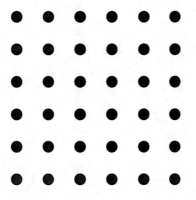

69

Geometry and the Coordinate Plane

A gift-wrap designer is working on a pattern for a box. She has created a template that will fold into a pyramid shape. She wants the box to have diagonal stripes on each side, starting at one bottom corner and wrapping all the way around the box. She has made a sketch, shown to the right, of the way she wants the box to look.

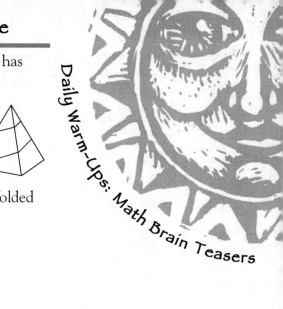

Help her out by drawing lines on the template so that the folded box will look like her sketch.

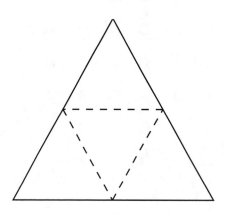

Geometry and the Coordinate Plane

A *tessellation* is a pattern made up of polygons that completely covers an area with no gaps or overlaps.

In the space below, create a tessellation with one of the following types of triangle:

- scalene

- isosceles

- equilateral

Geometry and the Coordinate Plane

A tricky mathematical pirate gave the map to his buried treasure to his nephew, Philippe. Help Philippe find the riches by following the directions. Then give the coordinates of the treasure.

1. Start at the palm tree. Go $(2{,}250 \div 450)$ paces north.

2. Go $(5.19 - 1.19)$ paces east.

3. Go (0.25×16) paces south.

4. Turn $\frac{1}{2}$ way around.

5. Go $(2.667 + 0.333)$ paces west.

6. Turn around 45°.

7. Go $(70.4 \div 35.2)$ paces north.

72

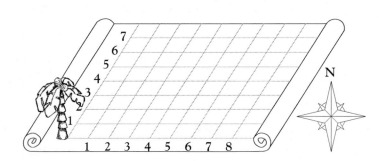

Geometry and the Coordinate Plane

Each numbered shape below is made up of five small squares. Can you fit them all into the 8 × 8 grid?

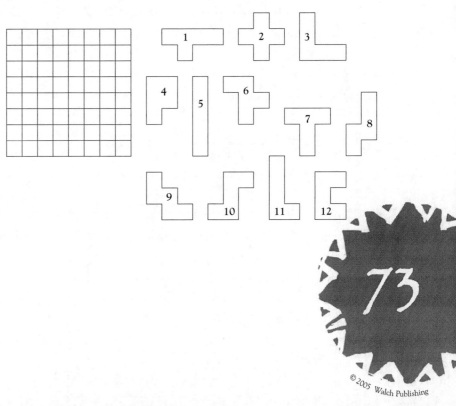

73

Geometry and the Coordinate Plane

Make a kite from the similar shapes below.

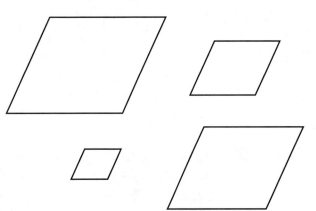

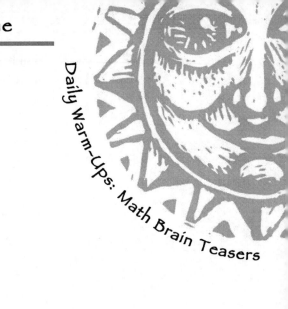

74

If the crank is turned clockwise, which way should the hamster run?

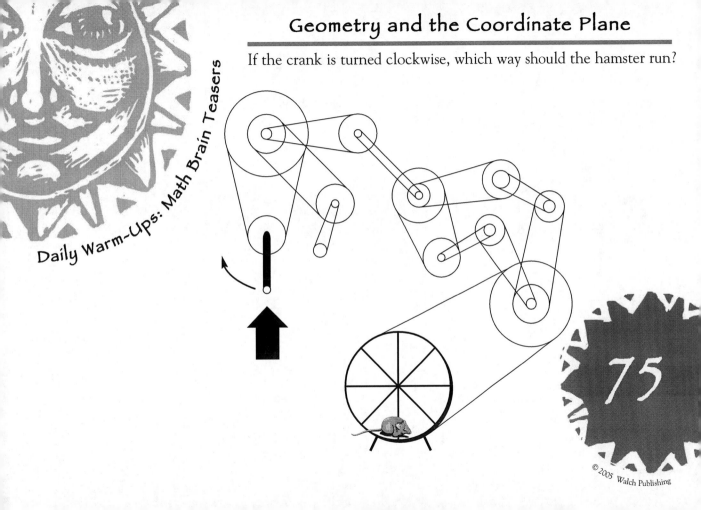

75

Money

The rules of a coin game state that a move is a jump over one or two coins at a time. What is the smallest number of moves it would take to get the following coins in the order P–P–P–N–N–N, with no gaps between coins?

76

A counterfeiter was trying to print his own money, and he came up with this version of a dollar bill. However, he has made a few mistakes. Can you find four errors on this fake bill?

77

Money

What is the smallest number of coins that can be combined to make 99 cents?

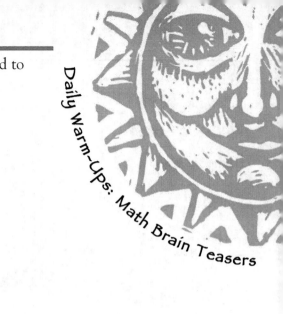

78

Money

You have $1.20 in change in your pocket, made up of 9 coins. There are 3 different types of coins, and 3 of each coin. What coins do you have?

Money

A swindler is playing a shell game. He has 5 pennies, 3 nickels, and 2 dimes. There are 5 cents under the first shell, 15 cents under the second shell, and 20 cents under the third shell. You do not know which coins are under each shell. You are allowed to see one coin at a time under the shell that holds 15 cents.

If you see a dime and a penny, can you tell which coins are under the three shells?

80

Money

A man is given 45 cents in change. He notices that his change includes just two kinds of coins. There are six coins in all. How many of each coin are there, and what coins are they?

Money

Draw lines to connect each profile to the coin on which it belongs.

Daily Warm-Ups: Math Brain Teasers

82

Money

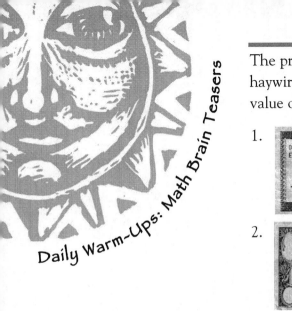

The printing press at the Department of the Treasury has gone haywire. Help the employees at the Treasury by writing the correct value on each bill.

1.

2.

3.

83

Money

There is a profile on the front of each of the following coins. What is the design on the back of each coin? (*Hint:* Do not include special-issue coins, such as state nickels and quarters.)

1. penny _____

2. nickel _____

3. dime _____

4. quarter _____

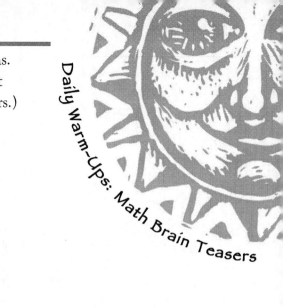

84

Money

You are a contestant on a show called Playing Games. You have to identify bills after only seeing part of the design on the back of each bill. See how observant you have been. Try to identify these bills from the designs on their backs.

1.

2.

3.

85

Money

Every January, Eva's parents decide what her weekly allowance will be for the coming year. This year, Eva has suggested a change. Instead of giving her the same amount every week, she suggests that her parents give her a varying amount. The first month, they can give her just 5 cents a week. Each month, they should double the previous month's allowance. Eva's parents accept her proposal.

If most months that year have 4 weeks each, and April, June, September, and November each have 5 weeks, how much allowance will Eva receive over the course of the year?

86

Money

A customer in a grocery store buys 2 for her young daughter. The cost is 30¢. She buys 65 for her mother. The cost is 60¢. Finally, she buys 101 for her grandmother. The cost is $1.20.

What is the customer buying?

87

Money

At the counter in a mini-market, there is a jar that says, "Give a penny, take a penny." There are 91 pennies in the jar at the start of the day. The first customer takes one penny from the jar. The next customer takes two pennies.

If this pattern continues, with each customer taking one penny more than the customer before, how many customers will it take to empty the jar?

88

Money

A nickel coin wrapper holds $2 worth of nickels. Liam has collected 240 nickels, which he plans to donate to charity.

How many nickel wrappers does he need?

89

Money

Little Red Riding Hood's grandmother has given her a big jar that she has been using to collect coins. Among the coins are 18 quarters. Her grandmother says that she can have all the coins if $\frac{1}{2}$ of the quarters go into her college savings account, $\frac{1}{3}$ of them go toward a new CD player, and $\frac{1}{9}$ go toward something fun. After happily agreeing to this, Little Red digs through the quarters and finds that one quarter is actually a subway token.

How many quarters should go in each category?

90

Money

Place 3 quarters, 3 dimes, and 3 nickels in 3 rows to form a square. Arrange them so that adding the coins in any horizontal row or vertical column gives a total of 40 cents.

91

Money

What is the smallest number of coins you can have so that subtracting the smaller coins from the largest one gives a result of 35 cents?

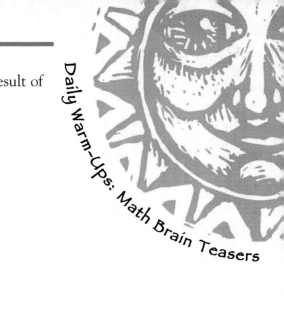

92

Money

Which would you choose: $100, or the total amount of pennies in a jar after 10 days, starting with one penny on day 1 and then adding double the preceding day's total amount each day?

Money

An ice-cream store calculates that each cone costs 75 cents in ingredients. A man walked into the store and bought an ice-cream cone for $2.50, paying with a $20 bill. After he had gone, the clerk realized that the bill he had given her was counterfeit. What is the total amount the store has lost?

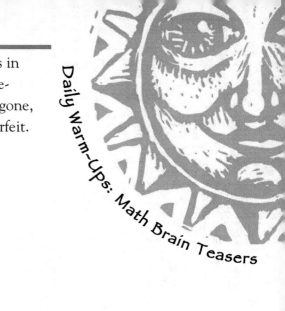

94

Money

A second-grader asked his mother for a quarter so he could buy a pretzel at lunch. His mother replied, "Only if you can spell it."

The second grader got the money. How?

95

© 2005 Walch Publishing

Money

In colonial times in the United States, money was based on the English system of pounds, shillings, and pence. The pound was the basic unit. It was divided into 20 shillings. A shilling was divided into 12 pennies, or pence. Pennies were divided into 4 farthings. How many farthings were there in a half pound?

96

Money

You have 12 coins. 11 of them weigh exactly the same, but 1 is heavier than the others. You want to find the heavier coin.

Using a balance scale, what is the smallest number of weighings you will need to find the heavy coin?

Daily Warm-Ups: Math Brain Teasers

Money

Bob had a car accident, and his car needed repairs. The repairs would take several days, so Bob needed to rent a car. Bob was able to rent a car for $24 a day. His insurance company offered to pay him $16 a day for every day he rented the car. However, they also offered to pay him $10 a day for every day he did *not* rent the car. Bob wants to rent the car for a few days and do without the car for a few days.

Is there any combination of rental days and non-rental days that will let Bob come out even—that is, receive exactly enough money from the insurance company to pay for the car rental?

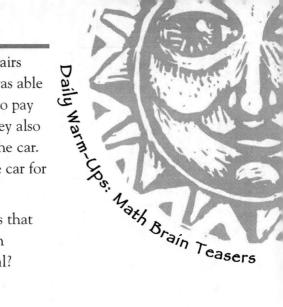

98

Find the name of the foreign coins below in this backward word find.

Location	Monetary Unit
Germany	euro
Israel	shekel
Japan	yen
Mexico	peso
Nigeria	naira
Switzerland	franc
United States	dollar

A	L	E	K	H	S
R	R	P	E	O	O
A	M	I	S	Y	R
L	L	R	A	E	U
L	L	M	Y	N	E
O	K	C	A	R	F
D	O	I	L	M	S

99

Daily Warm-Ups: Math Brain Teasers

Money

Louis gets paid $40 for digging a hole. He dug a hole for a client that was 5 feet wide by 3 feet deep by 2 feet long.

How much would he be paid for digging half a hole?

100

Money

Norton places a penny in his piggybank on July 1. Each day, Norton doubles the amount he places in the bank. On July 10, the piggybank is full.

On which day was the bank about half full?

Money

You have a nickel, a penny, and a quarter on your desk, in that order. How can you move the quarter to the right without touching either the quarter or the penny?

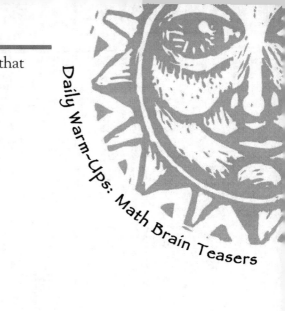

102

Money

Kenya has three coins that total 25 cents. One of the coins is not a dime. What are the other two coins?

103

Money

A very generous millionaire had $2.5 million. She gave away all but $400,000. She gave $50,000 to a children's burn unit, $175,000 to a homeless shelter, and another $175,000 for AIDS research.

After making all these donations, how much does she have left?

104

When will a one-dollar bill equal a five-dollar bill?

105

Measurement

Lance Legstrong cycles every day. His route is both circular and hilly. Each hill is as steep going up as it is coming down. When he leaves his driveway, he turns left. He travels up a hill, then down a hill, then along a straightaway. Then he goes up another hill, then there is another straightaway, then he goes up another hill, then down a hill.

Today, he decides to turn right as he leaves his driveway. Will his route be less strenuous?

106

Measurement

Which figure shows two pairs of rays that are of the same angle measurement?

a. 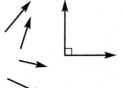   b. 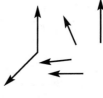   c.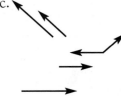

Measurement

What phrase does this picture suggest to you?

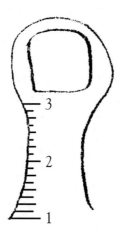

108

Measurement

Unscramble these jumbles to find some terms used in measurement.

1. MEEECITTNR _____

2. NIGFTHORT _____

3. DDECEA _____

4. ATNILROCMSOA TUIN _____

Measurement

Where does $1\frac{1}{2}$ plus $11\frac{1}{2}$ equal 1?

Measurement

How can you drop a lightbulb from a height of 4 feet and be certain it will not break?

111

Measurement

You may have heard the story of how Hannibal crossed the Alps with an army and a number of elephants. To move all his supplies, he must have used carts as well as elephants.

Imagine that some of Hannibal's carts had three wheels, one in the front and two in the back. For safety, each had a spare wheel, and the wheels were rotated so that all four wheels got the same amount of wear.

If Hannibal's entire route was 1,000 miles long, how many miles of wear were put on each wheel?

112

Measurement

What is the area of the shaded portion of the circle?

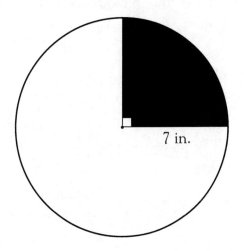

7 in.

Measurement

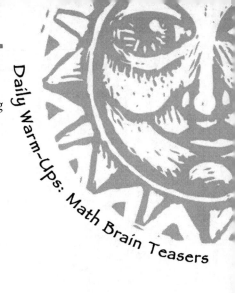

Berto's Buckets manufactures buckets in many different sizes. Berto is trying to describe all the sizes in a certain product line to a customer, but he has forgotten a few. The sizes increase according to a regular pattern.

Help Berto find the sizes of the missing buckets.

Size 1

1 quart

Size 2

1 gallon

Size 3

_____

Size 4

4 gallons

Size 5

_____

Size 6

9 gallons

114

Measurement

Art is an architect who has just finished drawing up plans for a television weather station. After reviewing the plans, Art's boss was puzzled by the angle shown below, and asked Art to double-check his measurements. Art realized he had made a mistake. Use a protractor to find out what the mistake was.

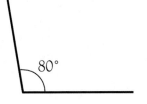

80°

115

Measurement

If you made a stack of nickels that reached to the moon, and a friend laid quarters end to end to reach the moon, whose coins would add up to the greatest value?

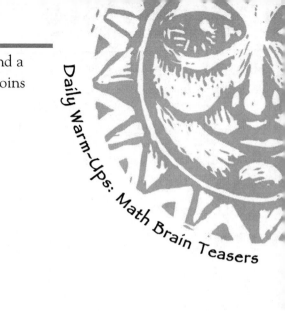

116

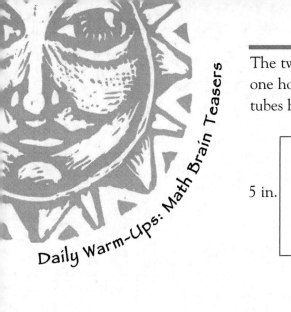

Measurement

The two note cards below have the same surface area. If you roll one horizontally and one vertically to make two tubes, will both tubes have the same volume?

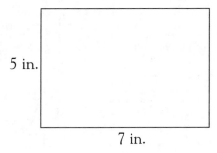

5 in.

7 in.

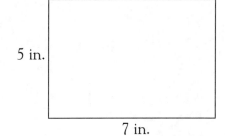

5 in.

7 in.

Daily Warm-Ups: Math Brain Teasers

Measurement

Some months have 30 days. Others have 31. How many have 28?

118

Measurement

What two U.S. bills still being circulated can be divided in half so that one person gets 999 times more than the other?

119

Measurement

At Ocean World, three dolphins—Spin, Span, and Spun—are performing flips out of the water. If Spin flips forward about 450°, Span flips backward about 540°, and Spun flips backward 270°, how do they appear when they enter the water?

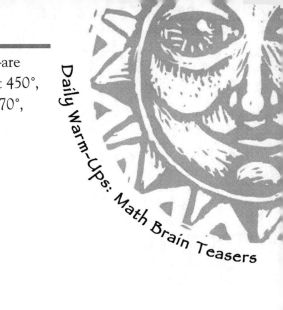

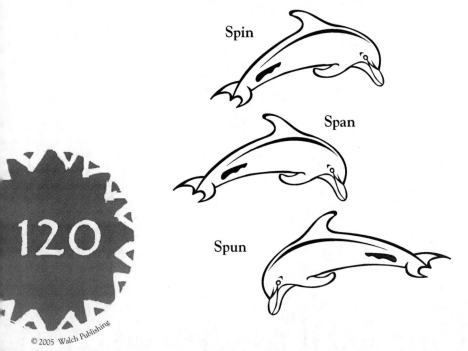

Spin

Span

Spun

120

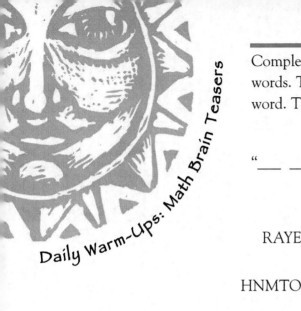

Complete the proverb by first unscrambling the letters to make words. Then use the letters in the circles to form another new word. This word completes the proverb.

" __ __ __ __ waits for no man."

RAYE

HNMTO

TYECNUR

ILNINMELUM

121

Daily Warm-Ups: Math Brain Teasers

Measurement

If you drop a 2-pound weight from a height of 2 feet into a 2-quart container of water that is at 45° C, and drop another 2-pound weight of the same mass and volume from the same height into another 2-quart container of water that is at 0° C, which weight will hit the bottom of the container first?

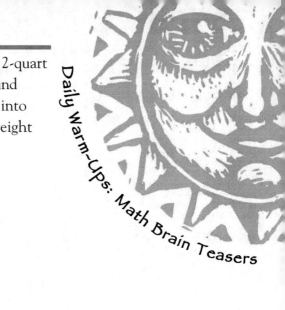

122

Measurement

The great detective Hercule Poirot looked closely at one angle in a triangle. He measured the angle and found that it measured 25°. Then he took out a magnifying glass that enlarged objects 5 times.

What would be the measure of the angle when viewed through the magnifying glass?

123

© 2005 Walch Publishing

Measurement

Earth weighs about 6 sextillion tons. The Great Wall of China is about 4,000 miles long. It ranges from 15 to 50 feet high, and from 15 to 30 feet wide.

If the wall is estimated to weigh about 1.5 sextillion tons and is destroyed by an earthquake, how much would Earth weigh?

124

Measurement

An egg-farming family were eating dinner. The younger daughter, who had been helping pack eggs for market, asked her father, "When does 4 plus 1 equal 60?"

Her father replied, "When 6 plus 2 equals 96."

Can you tell what they meant?

125

Measurement

Phillip was flying around the world in a straight line in his specially designed hot-air balloon. One day he realized that he had sailed 2 miles south, but he was now actually traveling north.

How can this be?

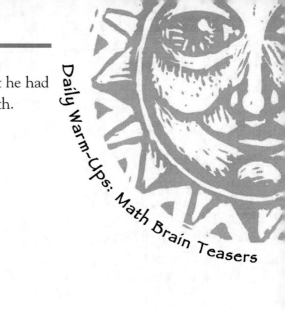

126

Looking in a store window, Jasper saw the reflection of the watch worn by the man standing next to him. Jasper thought, "Oh, I have plenty of time to get to my piano lesson. It does not start until 2:45."

Is Jasper correct?

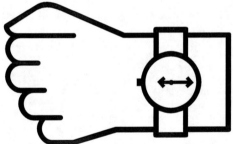

127

© 2005 Walch Publishing

Measurement

Ebenezer has two antique clocks. He sets both clocks to 1:00 P.M. Soon he realizes that one clock loses 10 seconds per minute.

When the correctly working clock shows 2:00 P.M., what will the malfunctioning clock show?

128

What is the perimeter of the net that could be folded to form this box?

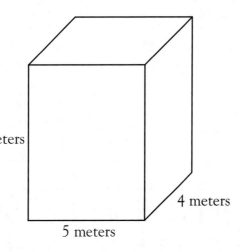

6 meters

4 meters

5 meters

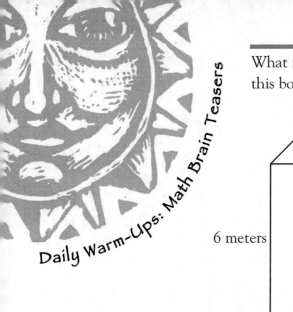

Measurement

Find the decimal word parts hidden in the puzzle below.

Meaning	Word Part	Meaning	Word Part
kilo-	1,000	deci-	$\frac{1}{10}$
hecto-	100	centi-	$\frac{1}{100}$
deka-	10	milli-	$\frac{1}{1,000}$

D	E	K	A	M	I	J	P	N	I	L
P	U	I	I	N	V	S	G	O	C	I
M	L	L	O	B	U	O	T	C	E	H
S	L	O	I	G	L	T	W	C	D	P
I	Z	D	F	E	C	E	N	T	I	G

Data Analysis, Statistics, Combinations, and Probability

All the students in a science class also take a math class. Half of the students who play baseball also take a math class. One-quarter of the students in the math class also take a science class. Sixty students play baseball. Forty students take a science class. Nobody who plays baseball takes a science class.

How many students who take a math class neither take a science class nor play baseball?

Data Analysis, Statistics, Combinations, and Probability

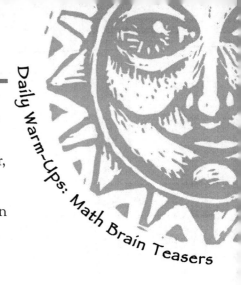

A student is given 11 quarters and 11 dimes. Her teacher tells her to place the coins in two bags. Once the coins are in the bags, the teacher will choose a coin from one of the bags. If the coin is a dime, the student gets to keep all the coins. If the coin is a quarter, the teacher takes back the coins.

The student wants to keep the coins. How can she put the coins in the bags to give herself the greatest chance of keeping them?

132

In a first-grade class, a teacher places the following cards on a table.

He then flips the cards over and mixes them up. He asks a student to pick up cards randomly and turn them over until she can make the word CAT.

What is the probability of the student making the word CAT on her first try without replacing any cards?

133

Data Analysis, Statistics, Combinations, and Probability

The writer of a math book wants a problem to read as follows:

"Which number in the data set could be eliminated without changing the mean?"

Give the writer a set of five numbers (they cannot all be the same number) that would make this problem work.

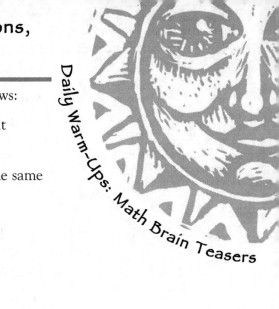

134

Data Analysis, Statistics, Combinations, and Probability

At the Happy family barbecue, the Happys are about to sit down at a picnic table. The family includes Grandma and Grandpa Happy, Father and Mother Happy, Ira Happy, and Iris Happy. None of the Happy females can stand sitting next to one another.

What is the number of possible combinations of Happys sitting around the picnic table?

135

Data Analysis, Statistics, Combinations, and Probability

Popcorn and cranberries are being strung in an alternating pattern (that is, pcpcpc . . .) for a winter holiday decoration. When the decoration is finished, it is tied off to make a circle. The finished decoration cannot include a piece of popcorn beside another piece of popcorn, or a cranberry beside another cranberry.

What numbers of popcorn pieces and cranberries can be used to make a string?

136

Assign each letter of the alphabet a number from 1 to 26 in order (that is, A = 1, B = 2 . . .). What is the range of the following phrase?

"Home, home on the range . . ."

137

Data Analysis, Statistics, Combinations, and Probability

Four cars are traveling on the road to No-Man's-Land, New Mexico. Two cars are going 45 miles per hour. The third car is going 57 miles per hour. The fourth is going 60 miles per hour.

Where is the median?

138

Data Analysis, Statistics, Combinations, and Probability

A middle school has just put new combination locks on all lockers. Each combination uses three numbers from 0 to 39. To open the lock, a student needs to turn the dial to each number in order, turning first to the right, then to the left, and then to the right again.

One student has dropped the piece of paper that his combination was written on. He is trying to decide which will be faster, going to the office to ask for another copy of the combination, or trying different combinations until he finds the correct one.

It will take him 15 minutes to get to the office and back. It will take him just three seconds to try a three-number combination. Which solution will be faster?

139

© 2005 Walch Publishing

Data Analysis, Statistics, Combinations, and Probability

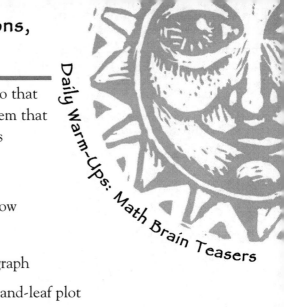

With a little imagination, you can write some math terms so that they show what they mean. For example, the *mode* is the item that occurs most often in a set of data. So, it might be written as

MMMMoodde

Think of a way to write the following terms so that they show their meanings.

box-and-whisker graph   line graph

bar graph   stem-and-leaf plot

140

Data Analysis, Statistics, Combinations, and Probability

Leo is teaching his brother Jem a new game. To play, both players roll a six-sided die so that two numbers are facing up. The smaller number is subtracted from the larger number. If the difference is 0, 1, or 2, player A gets 1 point. If the difference is 3, 4, or 5, player B gets 1 point. The game ends after 12 rounds. Since Leo is teaching Jem the game, he is player A.

After a few games, Jem starts to get suspicious. He thinks the game is rigged to let Leo win each time.

Is Jem correct? Or is the game fair?

141

Data Analysis, Statistics, Combinations, and Probability

At a birthday party, there are three flavors of ice cream: Fudge Ripple, Strawberry-Kiwi, and Cookie Dough Chunk. Each guest can have one or two flavors on a cone.

How many possible combinations are there?

142

Data Analysis, Statistics, Combinations, and Probability

All the high schools in Cumberland County have strong basketball teams. During basketball season, the local newspaper asks three residents to predict which team will win each game. Four games are scheduled for tonight. Here are today's predictions for the winners of each game.

Resident 1: Deering, Greeley, Cape, and Falmouth

Resident 2: Yarmouth, Scarborough, Cape, and Deering

Resident 3: Windham, Greeley, Scarborough, and Deering

No one chooses Portland.

Who is scheduled to play whom?

143

Data Analysis, Statistics, Combinations, and Probability

On average, in any group of 30 people, at least two people will have the same birthday.

For what month is this least likely to happen?

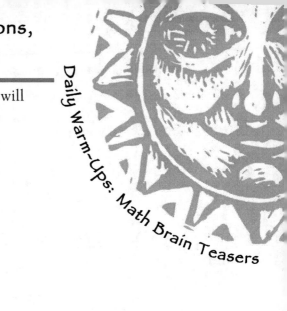

144

Data Analysis, Statistics, Combinations, and Probability

Use vowels to complete the words below.

1. T R _ _   D _ _ G R _ M

2. S _ M P L _   S _ Z _

3. F R _ Q _ _ N C Y

4. H _ S T _ G R _ P H

5. SC _ T T _ R   P L _ T

145

Data Analysis, Statistics, Combinations, and Probability

On his way out the door on a cold and snowy morning, Russ needed a pair of mittens. In a bin of mittens, there were 12 red mittens, 12 blue mittens, and 12 black mittens. All mittens can be worn on either the left hand or the right hand.

What is the greatest number of mittens Russ can pull out before he gets a pair of mittens that are the same color?

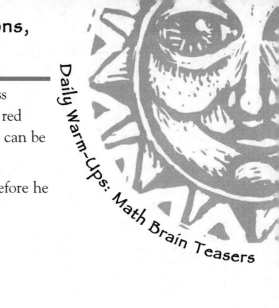

146

Data Analysis, Statistics, Combinations, and Probability

In genetics, some traits are dominant, and some are recessive. A dominant gene is shown with a capital letter. A recessive gene is shown with a small letter. If someone inherits both a dominant gene and a recessive gene for the same trait, the dominant gene will appear. The recessive gene only appears if the person inherits two recessive genes for the trait.

A woman with brown eyes (Bb) marries a man with blue eyes (bb). They decide to have children. The square shows the possible combinations of genes for eye color their children could receive.

	B	b	Mother
b	Bb	bb	
b	Bb	bb	

Father

If the woman and her husband have 3 children, what are the chances that all 3 children will have blue eyes?

147

Data Analysis, Statistics, Combinations, and Probability

Bonnie and Clyde are rolling two dice. Bonnie says to Clyde, "If the total on the dice is 6 or less, I'll wash the dishes. If the total on the dice is greater than 6, you must wash the dishes."

Should Clyde agree to this arrangement?

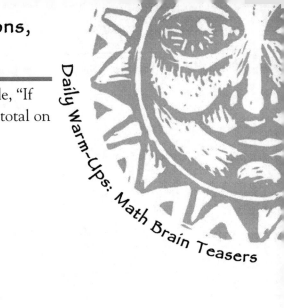

148

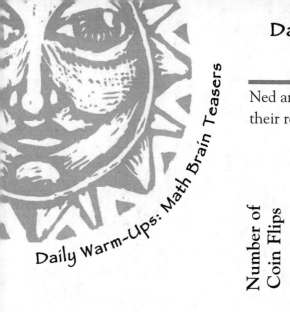

Ned and Harry have been flipping coins. The table below shows their results so far.

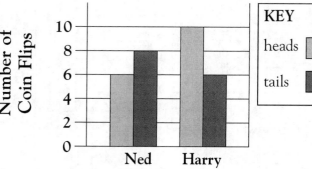

What are the chances of both Ned and Harry getting heads on the next flip of the coins?

Data Analysis, Statistics, Combinations, and Probability

The triangular pyramid has the letters C, U, B, E on its faces. The cube has the letters S, Q, U, A, R, E on its faces. If the shapes are rolled, what is the number of possible combinations if no combinations are repeated?

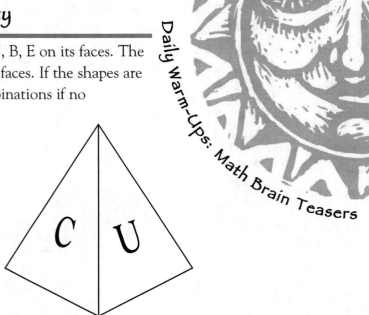
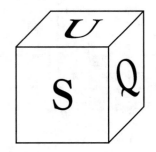

Data Analysis, Statistics, Combinations, and Probability

Hector has found himself at a crossroads in a lonely southwestern desert. He needs to call his uncle to ask in which direction he should travel to get to his uncle's house. Hector can remember his uncle's phone number, but he cannot seem to remember his three-digit area code.

If no zero appears as a first digit in any area code, what is the largest possible number of area codes Hector must try before he reaches his uncle?

Data Analysis, Statistics, Combinations, and Probability

Bubbles the clown has a bag containing bubble gum in three flavors: Bursting Berry Blast, Trouble Bubble, and Sinister Cinnamon. Gordon loves bubble gum and is eager to get his hands on some. Unfortunately, Bubbles says that he can only keep drawing gum from the bag until he has at least one of each flavor.

If Gordon reaches into the bag, what is the maximum number of pieces he can hope for before he has to stop drawing gum?

152

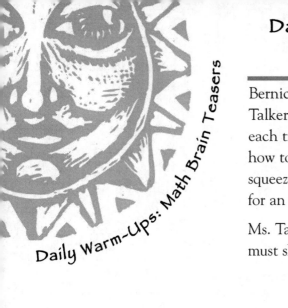

Data Analysis, Statistics, Combinations, and Probability

Bernice Talkalot is a contestant at the annual Chitty Chat Fast Talker Contest. Each contestant is given four 30-second trials. In each trial, contestants need to explain—as quickly as possible—how to make an apple pie. In her first three trials, Ms. Talkalot squeezed out 80, 85, and 89 words in 30 seconds. She had hoped for an average time of 90 words in 30 seconds.

Ms. Talkalot's fourth and last trial is coming up. How many words must she get in to average 90 words per trial?

153

Data Analysis, Statistics, Combinations, and Probability

Phil Anthropy is very generous. He has collected a pile of one-dollar bills to give equally to local charities. He knows that if he gives equal amounts to two charities he will have one dollar left over. The same happens if he gives the money to four charities. However, if he gives the money to five charities or to nine charities, he will have no money left over.

What is the minimum number of one-dollar bills in Phil's pile?

154

Data Analysis, Statistics, Combinations, and Probability

Euclid, the great mathematician, has locked all his important works of mathematics in his safe. Unfortunately, because Euclid is always busy remembering complex things, such as math theories and proofs, he forgets the simpler things—such as the four-digit combination to his safe. He remembers that the digits are 1, 2, 3, and 4, so he starts to write down possible combinations:

1234, 1134, 1114, 1111, 1214, 1222 . . .

How many possible four-digit combinations remain?

155

Fractions, Ratios, Decimals, and Percents

At a Halloween party, the guests voted for the costume they liked the best. The phantom was awarded the grand prize. The dragon received 15% of the votes. The vampire received $\frac{1}{10}$ of the vote, and the mermaid received one vote more than the vampire. The jack-in-the-box received 1 out of 20 votes. With $\frac{4}{20}$ of the votes, the rock star received more than the dragon but less than the phantom.

Place the costumes in order, from the most votes to the fewest votes.

156

Fractions, Ratios, Decimals, and Percents

At Far Back Away Farm, there is a fire in the barn. People rush to help put out the fire by getting in one of two lines of bucket brigades that are moving water from a stream to the barn. In one line, 6 people can put 160 gallons of water on the fire in 16 minutes. In the other line, 6 people can put 160 gallons of water on the fire in 20 minutes.

In all, how much water is put on the fire per minute?

157

© 2005 Walch Publishing

Fractions, Ratios, Decimals, and Percents

Four caterpillars decide to help one another encase themselves in cocoons. They encase one another in turn; once a caterpillar is fully encased, it can no longer help the next caterpillar.

If it takes one caterpillar two hours on its own to wrap a cocoon, how long will it be before all the caterpillars are in cocoons?

158

Fractions, Ratios, Decimals, and Percents

10 people can paint 60 houses in 120 days. How many days will it take 5 people to paint 30 houses?

Daily Warm-Ups: Math Brain Teasers

159

Fractions, Ratios, Decimals, and Percents

Draw four straight lines to divide the rectangle shown below into seven pieces.

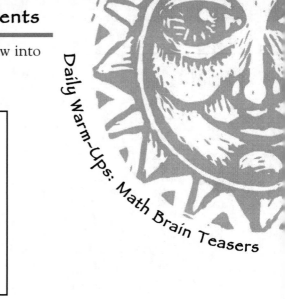

Fractions, Ratios, Decimals, and Percents

Juliet is angry with Romeo, and she decides to sit on the far side of the sofa away from Romeo while they watch television. Romeo, trying to make up with her, moves one half the distance of the sofa toward Juliet and looks for some sort of forgiveness in her eyes. He then moves one half that distance again toward his beloved. If Romeo continues this pattern, does he ever reach Juliet?

161

Fractions, Ratios, Decimals, and Percents

Hooray! You have just won the lottery. The state lottery commission says that you can have one hundred-thousandth of a billion dollars, one hundred million ten-millionths of a dollar, or one million millionths of a dollar.

What do you choose?

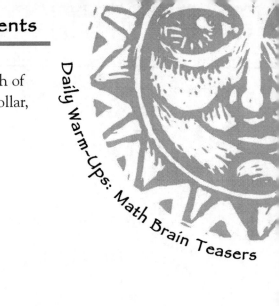

162

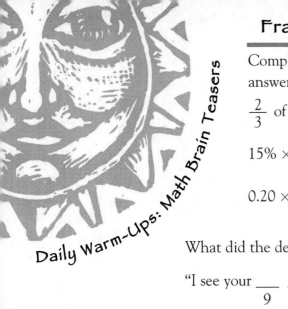

Fractions, Ratios, Decimals, and Percents

Complete the following calculations. Then use the letters of each answer to finish the riddle.

$\frac{2}{3}$ of $\frac{3}{2}$ = _____ (i)   40% of 25 = _____ (s)

15% × 20 = _____ (n) $\frac{3}{4}$ of 12 = _____ (d)

0.20 × 40 = _____ (v)   50% of 22 = _____ (o)

What did the decimal say to the fraction?

"I see your ___ ___ ___ ___ ___ ___ ___ ___."
          9   1   8   1   10   1   11   3

163

Fractions, Ratios, Decimals, and Percents

If it takes an Amigo Airlines' DC-10 5 hours and 25 minutes to fly from New York to Los Angeles, how long would it take 3 DC-10s to fly from New York to Los Angeles?

164

What is three halves, divided by three halves, divided by four thirds, divided by five quarters, divided by six fifths, divided by six sevenths?

165

© 2005 Walch Publishing

Fractions, Ratios, Decimals, and Percents

0.2^{100} is a very small number when it is written in standard form. If this number were written in standard form, what would be the last digit to the right?

166

Fractions, Ratios, Decimals, and Percents

How good a mathematical detective are you? There are two possible solutions to the problem below. Can you find them both? Here are the clues.

$$\frac{8}{?} + \frac{?}{6} = \frac{1}{2}$$

$$\frac{8}{?} + \frac{?}{6} = \frac{1}{2}$$

167

© 2005 Walch Publishing

Fractions, Ratios, Decimals, and Percents

Pedro's pogo stick store is going out of business. Pedro is offering Big Bounce pogo sticks for 70% off the original price. He is selling Boing-Boing pogo sticks, which have already been reduced in price by 50%, for another 20% off.

If the original cost for both sticks was $30, do the reduced-price pogo sticks cost the same amount as each other?

168

Fractions, Ratios, Decimals, and Percents

John Doe's average heart rate is about 70 beats per minute. If you start counting heartbeats right after midnight on January 1, on what date will his heart have beaten 1,000,000 times?

169

Fractions, Ratios, Decimals, and Percents

When Cassie Pemberton was one year old, the ratio of her age to her mother's was 1 to 35. Today, the ratio is 1 to 2.

How old is Cassie now?

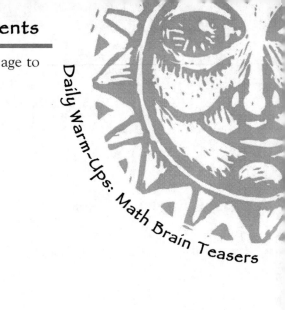

170

Fractions, Ratios, Decimals, and Percents

Look closely at the equiangular equilateral triangle. Points A, B, and C are midpoints. What is the ratio of the area of the large triangle to the area of the smaller, shaded right triangle?

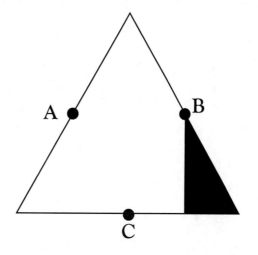

171

© 2005 Walch Publishing

Fractions, Ratios, Decimals, and Percents

A tycoon left one half of his fortune to his wife, a quarter of his fortune to his only daughter, and one fifth of his fortune to a sister. The rest of his estate went to his beloved Irish wolfhound, Con.

If the tycoon left 5 million dollars to his dog, how much money did his wife inherit?

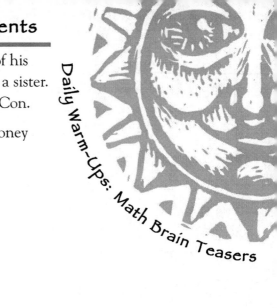

172

Fractions, Ratios, Decimals, and Percents

Lester, an observational scientist, was doing an experiment with fleas. He wanted to know if playing soft music would put fleas to sleep. Lester set up 5 tanks, with 100 fleas in each tank. He then played music for the fleas. After playing the music, Lester made the following observations: tank 1—10% asleep; tank 2—70% asleep; tank 3—70% asleep; tank 4—70% asleep; tank 5—20% asleep.

His conclusion was that $\frac{3}{5}$ of the time, playing soft music made 70% of the fleas fall asleep. Was this conclusion correct?

173

Fractions, Ratios, Decimals, and Percents

The diameter of a smaller wheel on a bicycle is $\frac{7}{22}$ that of the larger wheel. How many times larger is the circumference of the larger wheel than that of the smaller wheel?

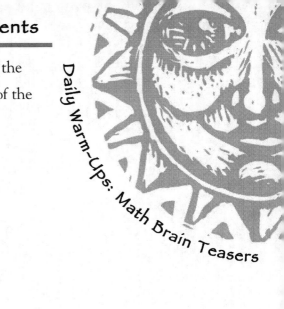

174

Fractions, Ratios, Decimals, and Percents

There are 3 gears. The circumference of the smallest gear is $\frac{1}{2}$ that of the largest gear. The circumference of the middle gear is $\frac{2}{3}$ that of the largest gear.

How many times larger is the circumference of the middle gear than that of the smallest gear?

Fractions, Ratios, Decimals, and Percents

Professor Camlink, Professor of Engineering at Cogs and Gears University, has just noticed something interesting about the three gears below. For every 1 full revolution of the largest gear, the smallest gear goes around 3 times. For every 1 full revolution of the middle gear, the smallest gear goes around $1\frac{1}{2}$ times.

Help the professor use this information to tell how many more times than the largest gear the middle gear will turn every time the smallest gear makes 1 full revolution.

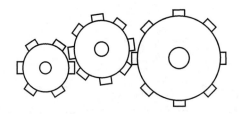

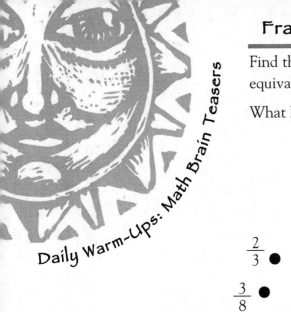

Fractions, Ratios, Decimals, and Percents

Find the answer to the riddle by drawing lines to connect equivalent numbers.

What kind of prism does the U.S. Army like? _____

$\frac{7}{8}$ ●   0.375 ●

$\frac{2}{3}$ ●   ● 2%

$\frac{3}{8}$ ●   ● $87\frac{1}{2}$ %

0.2 ●   ● $\frac{1}{5}$

$66\frac{2}{3}$% ●   ● 0.02

Daily Warm-Ups: Math Brain Teasers

Fractions, Ratios, Decimals, and Percents

What is the next number in the sequence?

$9.\overline{09}\%$, $\dfrac{2}{11}$, $0.\overline{27}$, $36.\overline{36}\%$, _____

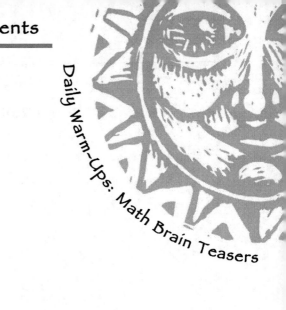

178

Fractions, Ratios, Decimals, and Percents

A slightly clumsy math teacher, Mr. Ichabod Crane, tripped over his shoelace, and the cards he was carrying fell on the floor. Help him place these cards in order, from greatest to least.

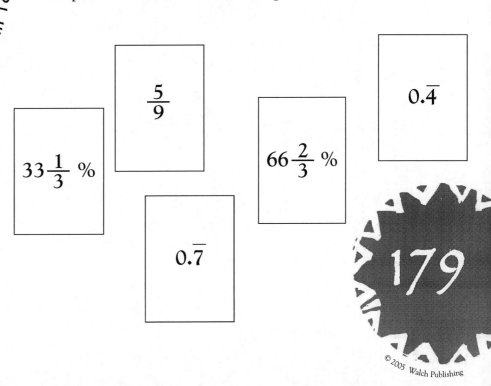

$33\dfrac{1}{3}$ %

$\dfrac{5}{9}$

$0.\overline{7}$

$66\dfrac{2}{3}$ %

$0.\overline{4}$

179

Fractions, Ratios, Decimals, and Percents

Which three fractions can be added for a sum of 1?

$$\frac{1}{2} \qquad \frac{1}{3} \qquad \frac{1}{4} \qquad \frac{1}{5} \qquad \frac{1}{6}$$

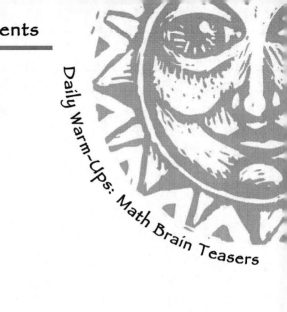

180

Numbers, Numeration, Operations, and Patterns

1. Most combinations of 0, 1, 8 Sample answer: .11, mirror image 11

2. 1,055,555 (first swipe—10, second swipe—100 + 5, third swipe,—1,000 + 50 + 5, fourth swipe—10,000 + 500 + 50 + 5, fifth swipe—100,000 + 5,000 + 500 + 50 + 5, sixth swipe—1,000,000 + 50,000 + 5,000 + 500 + 50 + 5)

3. 32 (3750 ÷ 10 = 375 ÷ 5 = 75 ÷ 3 = 25 + 7 = 32)

4. Any number between 2,050 and 2,149

5. Anya should choose the second plan. (With the first payment plan, she would earn a total of $140 after seven days (7 × 20 = 140). With the second plan, she would earn 2 + 4 + 8 + 16 + 32 + 64 + 128, or $254.)

6. Answers will vary. Sample answer: 9 = 2 + 7, put the digits beside each other to make 27.

7. 89 (Each number in the Fibonacci sequence is made by adding together the last two numbers in the sequence.)

8.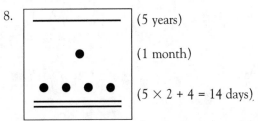

9. Rat (In 60 years, 5 cycles will have passed, and the year will be 2065. It will be the year of the rooster. Three years later, in 2068, it will be the year of the rat.)

10. Great-aunt (The only choices are an aunt or a grandmother. Since chick 602 is not in a vertical line with chick 2, chick 2 must be an aunt. From the number of generations between the chicks, chick 2 is a great-aunt.)

Daily Warm-Ups: Math Brain Teasers

11.

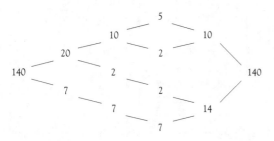

12. Two palindromic combinations of the digits 1 through 4 added together will make a palindrome. Sample answer: 1441 + 2332 = 3773

13. Carmen needs 20 packages of hot dogs and 25 packages of buns for a total of 200 of each item.

14. Melinda: –40, Melana: 100, Melony: 60, Melody: –10

15. 25.85 + .190 + 46.31 = 72.35

16. A shoe (Mallard has completed 5 cycles of chewing on the items and has begun the sixth cycle. He is chewing on the second item, which is a shoe.)

17.
```
    8 3 2
  × 4 6 4
    3 3 2 8
  4 9 9 2 0
  3 3 2 8 0 0
  3 8 6 0 4 8
```

18. The greatest number possible is 10.

19. 19,900 (1 + 199 = 200, 2 + 198 = 200, . . . 101 + 99 = 200; 200 × 99 + 100 = 19,900)

20. 28 (Factors of 28 (other than 28 itself) are 1, 2, 4, 7, and 14, which add up to 28.)

21. .55 ((18 – 16.9) ÷ 5 × 2.5)

22. 20 (When Silas is 4, Opal is 4 × 3, or 12; the difference in their ages is 12 – 4 = 8. When Silas is 12, Opal will be 12 + 8, or 20.)

23. .45 and 28.035, 4.5 and 280.35, 45 and 2803.5

24. 10 turns (additional rocks on the beach after each turn: 1, 2, 3, 4, 5, 6, 7, 8, 9, and 10; number of rocks on the beach: 1 + 2 + 3 + 4 + 5 + 6 + 7 + 8 + 9 + 10 = 55)

25. About 32 years (There are $60 \times 60 \times 24 \times 365 =$ 31,536,000 seconds in a year, so a billion seconds would make 31.71 years.)
26. Answers will vary.
27. Square roots
28. 2 (Each horizontal row adds up to 1 more than the row above. So, the last row must add up to 12.)
29. 55 (Using a table, a pattern emerges.)

Trees	1	2	3	4	5	6	7	8	9	10	11
Tangles	0	1	3	6	10	15	21	28	36	45	55

30. $144 + 256 = 400$ ($12^2 + 16^2 = 20^2$)
31. 8 ($-3, -4, -5$ and $3, 4, 5$; $-3, -4, 5$ and $3, 4, -5$; $-3, 4, 5$ and $3, -4, -5$; $-3, 4, -5$ and $3, -4, 5$; $3, -4, 5$ and $-3, 4, -5$; $3, 4, 5$ and $-3, -4, -5$; $3, 4, -5$ and $-3, -4, 5$)
32. 2^0 and 3^0 (If 2 and 3 are raised to the zero power, they will then equal each other because any number raised to the zero power is 1.)

33. $39 + 1 = 40$ and $31 + 9 = 40$ (Making a chart of all the combinations of numbers and adding each combinations to the other combinations will find the necessary digits.)

	0	1	3	4	9
0		01	03	04	09
1	01		13	14	19
3	30	31		34	39
4	04	41	43		49
9	09	91	93	94	

34. MATH IS COOL ($\sqrt{169} = 13$, M; $15 + -14 = 1$, A; $4 \times 5 = 20$, T; $-2 \times -4 = 8$, H; $3^2 = 9$, I; $33 - 14 = 19$, S; $-75 \div -25 = 3$, C; $5.6 + 9.4 = 15$, O; $33 \div 2.2 = 15$, O; $16.6 - 4.6 = 12$, L)

Daily Warm-Ups: Math Brain Teasers

35.

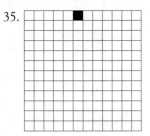

Geometry and the Coordinate Plane

36.

37. 7 (H Θ I Ξ O ΦΧ)

38.

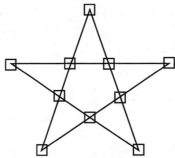

39. Check students' answers.

40. 32 (The 9th person is opposite the 25th. Since 25 − 9 = 16, there are 16 people in half the circle, so there are 32 people in the whole circle.)

41.

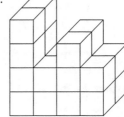

42.

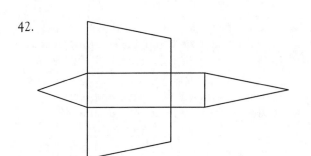

43.

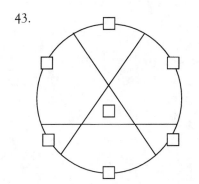

44. 1. Square (The relationship is that the second word is a type of the first.)

2. 7 (The relationship is that the first word looks like a number.)

45.

(Each symbol shows a digit and its mirror image. The next number in the series is 7.)

46. Rectangular covers could fall through the holes. (Because the diagonal of a rectangle is always longer than either of its sides, a rectangular cover could be dropped through the hole if it was dropped at an angle, despite the protruding lip. However, because a circle is the same width all around, a circular cover cannot fall through the hole, because the lip makes the hole smaller than any part of the circle.)

47. 25 feet (fraction of pole in the ground: $\frac{1}{10}$;

    fraction covered by water: $\frac{1}{2}$; $\frac{1}{2} + \frac{1}{10} =$

    $\frac{(5 + 1)}{10} = \frac{6}{10} = \frac{3}{5}$. Fraction of pole out of water

    $= 1 - \frac{3}{5} = \frac{2}{5}$. Thus, $\frac{2}{5}$ of the pole is 10 feet,

    and the total length of the pole is 25 feet.)

48. 7

49. The spiders will get to their flies at the same time.
    (The distance from spider A to fly A is the radius
    of the web. Spider B to fly B is a diagonal of the
    rectangle, which matches the diagonal that is also
    a radius.)

50. 17 feet (The angle between the telephone pole
    and the ground is 90°, and if one of the angles is
    45°, the other angle must also be 45°. This is an
    isosceles right triangle. Thus, two of the sides are
    equivalent in length. Use the Pythagorean
    theorem to find the length of the third side.)

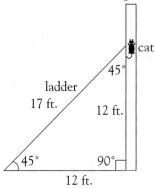

51. 300 sq cm (It does not matter that the shapes are puzzle pieces. There are 12 pieces: 4 corner pieces, 6 pieces with only one edge, and 2 interior pieces. This means that the puzzle is 3 5-centimeter pieces by 4 5-centimeter pieces. So, the area is (3 × 5) × (4 × 5) = 300 square centimeters.)

52. 44 sq units

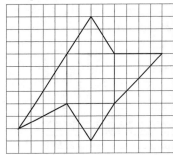

(Break the irregular form into recognizable geometric shapes, and use the appropriate formulas to find the area of each shape. For triangles, $A = \frac{1}{2} bh$, and for squares, $A = s^2$. Adding the areas gives 6 + 8 + 6 + 8 + 16 = 44 square units.)

53. 4,624 sq ft (Working backward, find the area of the ninth square, then the eighth, then the seventh, and so forth. A chart appears below.

Time around	Dimension to mow
ninth	4 by 4
eighth	12 by 12
seventh	20 by 20
sixth	28 by 28
fifth	36 by 36
fourth	44 by 44
third	52 by 52
second	60 by 60
first	68 by 68

Finally, the dimensions of the 1st time around gives the dimensions of the lawn, 68 feet long by 68 feet wide. So, the area is 68 × 68 = 4,624 sq ft.)

54.

55.

56.

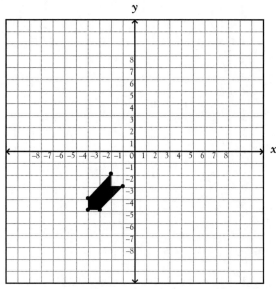

57. The only similar figures are the two squares.

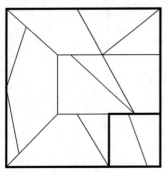

58. perimeter ≈ 25.12 units (The figure is divided into four semicircles and one rectangle. The lengths of the sides of the rectangle are the diameters of the semicircle. The circumference of each figure is found and added together. Using 3.14 as a value for π, the perimeter = (2 × 3.14) + (6 × 3.14) ≈ 15.12.)

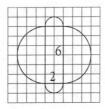

59.   30°   parallel lines

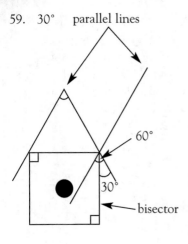

60.   102.5°

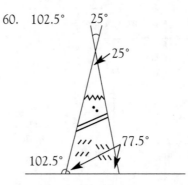

61.   "Complementary" angles
62.   A striped, horizontal cylinder
63.   4 × 4
64.   d (a. square—area = 8 square units, perimeter = 12 units; b. right triangle—area = 6 square units, perimeter = 12 units; c. parallelogram—area = 28 square units, perimeter = 30 units; d. trapezoid—area = 108.4 square units, perimeter = 42 units)

65.

66. 432 ice cubes (The volume of the ice chest is
$12 \times 12 \times 24 = 3{,}456$ sq in. The volume of an ice
cube is $2 \times 2 \times 2 = 8$ sq in.
$3{,}456$ sq in. $\div 8$ sq in. $= 432$ ice cubes.)

67. 9.42 square units (The area of the outer circle is
$\pi \times 2 \times 2 \approx 12.56$ square units. The area of the
pupil is $\pi \times 1 \times 1 \approx 3.14$ square units. The area
of the iris is $12.56 - 3.14 \approx 9.42$ square units.)

68.

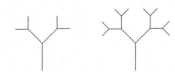

69. Answers will vary. Here is one solution.

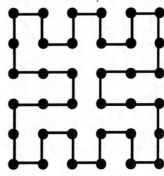

70.

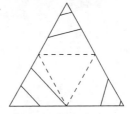

71. Check students' answers. (Be sure there are no
gaps or overlaps.)

Daily Warm-Ups: Math Brain Teasers

72. (1, 3)

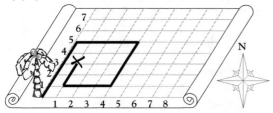

(The turns have nothing to do with the plotting the points. NSEW do not change on a map.)

73.

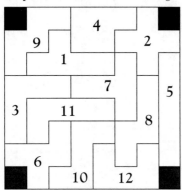

74. Answers may vary.

75. Clockwise

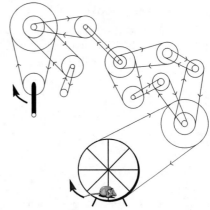

Daily Warm-Ups: Math Brain Teasers

Money

76. 6 (One solution: move 1: np_pnpn; move 2: npppn_n; move 3: npp_npn; move 4: ppnnpn_; move 5: pp_nnpn; move 6: pppnnn)

77. George Washington is facing left. He is smiling. TEN is written where ONE should be written. It should read "THIS NOTE IS LEGAL TENDER FOR ALL DEBTS, PUBLIC AND PRIVATE."

78. 8 (1 half-dollar, 1 quarter, 2 dimes, 4 pennies)

79. 3 quarters, 3 dimes, and 3 nickels

80. Yes (The table shows all the possible combinations. If the 15-cent shell includes a dime and a penny, then the last combination in the table must be the actual one.)

5¢	15¢	20¢
5 pennies	3 nickels	2 dimes
5 pennies	1 dime and 1 nickel	1 dime and 2 nickels
1 nickel	2 nickels and 5 pennies	2 dimes
1 nickel	1 nickel and 1 dime	1 dime, 1 nickel, and 5 pennies
1 nickel	1 dime and 5 pennies	1 dime and 2 nickels

81. 3 dimes and 3 nickels

82.

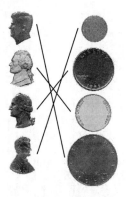

83. 1. $5, 2. $1, 3. $10
84. 1. Lincoln Memorial; 2. Monticello; 3. a torch, an olive branch, and an oak branch; 4. an eagle
85. 1. $10, 2. $5, 3. $1

86. $885 (Starting with .05 per week in January and doubling the weekly amount every month, Eva will receive $(.05 \times 4) + (.10 \times 4) + (.20 \times 4) + (.40 \times 5) + (.80 \times 4) + (1.60 \times 5) + (3.20 \times 4) + (6.40 \times 4) + (12.80 \times 5) + (25.60 \times 4) + (51.20 \times 5) + (102.40 \times 4) = .20 + .40 + .80 + 2.00 + 3.20 + 8.00 + 12.80 + 25.60 + 64.00 + 102.40 + 256.00 + 409.60 = 885.00$)

87. Number-shaped birthday candles
88. 13 $(1 + 2 + 3 + 4 + 5 + 6 + 7 + 8 + 9 + 10 + 11 + 12 + 13 = 91)$
89. 6 ($2 ÷ 0.05 = 40 nickels in a wrapper. 240 nickels ÷ 40 = 6 filled wrappers.)
90. 9, 6, and 2 quarters (The total of the quarters plus the token is used. 18 quarters $\times \frac{1}{2}$ = 9 quarters, 9 go into the college savings account. $18 \times \frac{1}{3}$ = 6 quarters, 3 quarters go toward a CD player. Finally, $18 \times \frac{1}{9}$ = 2 quarters, 2 quarters are spent on something fun. The remaining coin is the token.)

91. Arrange the rows and columns so that there is a quarter, a dime, and a nickel in each.

    n q d
    d n q
    q d n

92. 3 (half-dollar, dime, nickel)

93. The jar of pennies (D1: 1; D2: 1 + (2 × 1) = 3; D3: 3 + (2 × 3) = 9; D4: 9 + (2 × 9) = 27; D5: 27 + (2 × 27) = 81; D6: 81 + (2 × 81) = 243; D7: 243 + (2 × 243) = 729; D8: 729 + (2 × 729) = 2187; D9: 2187 + (2 × 2187) = 6561; D10: 6561 + (2 × 6561) = 19683 cents = $196.83)

94. $18.25 ($0.75 + $20.00 − $2.50)

95. He spelled "it"—I T.

96. 480 (a half pound = 10 shillings, 12 pence × 10 shillings = 120 pence, 4 farthings × 120 pence = 480 farthings)

97. 3 (First weighing: divide the coins into 2 groups of 6, then weigh them against each other; the heavier group contains the heavy coin. Second weighing: divide the 6 coins from the heavier group into 2 groups of 3, weigh them against each other; the heavier group contains the heavy coin. Third weighing: place 1 of the 3 coins from the heavier group in each pan of the scales. If the coins balance, then the third coin in that group is the heavy one. If the coins do not balance, the one what weighs more is the heavy one.)

98. 4 days with no rental and 5 days with a rental

Day	1	2	3	4	5	6	7	8	9
No Rental	+$10	+$10	+$10	+$10					
Rental					−$8	−$8	−$8	−$8	−$8

Daily Warm-Ups: Math Brain Teasers

99.

100. $40 (Half of a hole is still a hole.)
101. July 9

Day	1	2	3	4	5	6	7	8	9	10
# of pennies	1	2	4	8	16	32	64	128	256	512

(The total, after day 10, is 1,023 pennies. Half this amount is 511.5 which is the total after July 9.)

102. Place a finger on the nickel. Slide it to the left, then quickly move it to the right, hitting the dime. The dime slides into the quarter, which moves.

103. Dimes (One coin is not a dime, but it must be less than a quarter. Since the three coins must add up to 25 cents, it must be more than a penny. So, it must be a nickel, which means the other two coins must be dimes.)
104. $400,000
105. A one-dollar bill will equal a five-dollar bill when you examine any of their physical attributes, such as height, width, area, or perimeter.

Measurement
106. No (He will cover the same terrain, just in reverse order.)
107. Figure c (Although the last pair of rays in figure c are reversed, this is the only set of rays that can form the same angle.)
108. A rule of thumb
109. 1. centimeter, 2. fortnight, 3. decade, 4. astronomical unit
110. On a clock ($11\frac{1}{2}$ hours added to 1:30 is 1:00.)

Daily Warm-Ups: Math Brain Teasers

111. Drop it from a height greater than 4 feet. (It will drop 4 feet without breaking, and will only break when it hits the floor after dropping more than 4 feet.)

112. 750 miles (Since the four wheels of the three-wheeled cart share the journey equally, each wheel covers three fourths of the total distance. $1000 \times .75 = 750$.)

113. 38.5 square inches ($90/360 \times \pi \times r^2$, $\frac{1}{4} \times 3.14 \times 49 = 38.5$ square inches)

114. 2 gallons + 1 quart, 6 gallons + 1 quart (The number of the bucket is squared to give the number of quarts, and then the number of quarts is simplified to find the number of gallons.)

Bucket	1	2	3	4	5	6
Number of quarts	1	4	9	16	25	36
Sizes	1 quart	1 gallon	2 gallons + 1 quart	4 gallons	6 gallons + 1 quart	9 gallons

115. Art placed the protractor on the angle backward and read the angle measurement as 80°.

116. The stack of nickels will be of a greater value. (It takes between 12 and 13 nickels to equal the diameter of one quarter. So, for every 25 cents in the quarter stack, the nickel stack will have a value of 60 to 65 cents.)

117. The tube that was made from being rolled horizontally—the shorter tube—will have a greater volume. (Using 5-by-7 note cards, two tubes are made. The vertical tube has a height of 7 and bases of circumference 5. $5 = 3.14 \times d$, $5/3.14 \approx 1.59$, $r = 1.59/2 \approx .79$; the volume of the tube is $\pi r^2 h = 3.14 (.79)^2 \times 7 \approx 13.71$ inches^3. The horizontal tube has a height of 5 and bases of circumference 7. $7 = 3.14 \times d = 7/3.14 \approx 2.23$, $r = 2.23/2 \approx 1.11$; the volume of the tube is $\pi r^2 h = 3.14 (1.11)^2 \times 5 \approx 19.34$ inches^3.)

118. All months have 28 days.

119. $1 and $1000

Daily Warm-Ups: Math Brain Teasers

120.

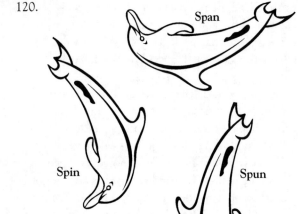

Span

Spin

Spun

121. Time (The words are **y**ear, **m**onth, century, millennium. Unscrambling e, m, t, i spells the word *time*.)

122. The weight that was dropped into the container at 45° C would hit the bottom first. (The water in the second container would be frozen solid.)

123. 25° (The lines that make up the angle may appear larger, but so does the space between them; the angle itself does not change in size.)

124. 6 sextillion tons (The Great Wall of China might have been turned into rubble, but the total weight of the rubble has not changed.)

125. They were referring to dozens of eggs. (4 dozen plus 1 dozen equals 5 dozen, or 60; 6 dozen plus 2 dozen equals 8 dozen, or 96)

126. Phillip was traveling toward the South Pole in a straight line. He traveled 1 mile south to the South Pole, crossed over the Pole, and then, still maintaining his straight line, began going north.

127. No (Using the stem on the watch to determine the top of the watch, the time is actually 2:45.)

128. 1:50 P.M. (The malfunctioning clock loses 10 seconds per minute, and there are 60 minutes in one hour, so the clock has lost $10 \times 60 = 600$ seconds, or 10 minutes.)

129. 74 meters

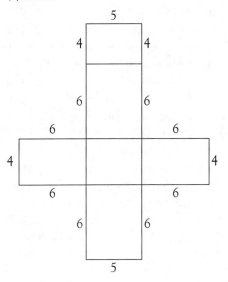

130.

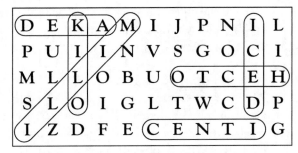

Data Analysis, Statistics, Combinations, and Probability

131. 90 students (Science—40 math students, 0 play baseball; baseball—30 math students, 0 science students; math—40 science students = $\frac{1}{4}$ math class; $40 \times 4 = 160$ math students; 30 play baseball, no science students play baseball. So 160 students – 40 science students – 30 students who play baseball = 90 math students who neither study science nor play baseball.)

Daily Warm-Ups: Math Brain Teasers

132. If she places all the quarters in one bag and all the dimes in the other, the probability of the teacher drawing a dime is 1/2.

133. 1/21 (P(C) = 2/9, P(A) = 4/8, P(T) = 3/7; P(CAT) = 2/9 × 4/8 × 3/7 = 24/504 = 1/21)

134. Answers will vary. Sample answer: 0, 1, 2, 1, 2

135. 72 (The first seat can be filled by any 1 of the 6 family members. Alternating sexes, the next seat can be filled by 1 of the 3 family members of the opposite sex, the third seat must be 1 of 2, the fourth seat must be 1 of 2, and the fifth and sixth seats have only one option each. 6 × 3 × 2 × 2 × 1 × 1 = 72.)

136. All the even numbers, with the same number for both items.

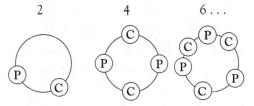

137. 19 (a = 1 and t = 20; the range is 20 − 1 = 19)

138. The median is in the middle of the road.

139. Going back to the office (With 40 numbers to choose from, and three numbers in each combination, there are 64,000 possible combinations for each lock (40 × 40 × 40 = 64,000). If each attempt takes 3 seconds, it will take him a total of 192,000 seconds to try all 64,000 combinations. That is over 53 hours. Unless he strikes it lucky and gets the correct combination on one of his first 300 tries, going back to the office for another copy of the combination will be faster.)

140. Answers will vary.

141. Jem is right; the game is not fair. There are 36 possible outcomes for rolling two dice. For 24 of them, the difference between the lower and higher number will be 0, 1, or 2. Only 12 of them can have a difference of 3, 4, or 5. Player A is twice as likely as Player B to win a point in each round.

142. 9 combinations (3 single cones, 6 double cones when repeated combinations are subtracted)

143. Deering will play Portland, Greeley will play Yarmouth, Cape will play Windham, and Falmouth will play Scarborough.

144. February (It has at most 29 days.)

145. 1. tree diagram, 2. sample size, 3. frequency, 4. histograph, 5. scatter plot

146. 4 mittens (There are 3 colors. Even if the first 3 mittens are all of different colors, the fourth mitten must match one of the first 3.)

147. 1/8 (Each child has a probability of 1/2 of having blue eyes; the probability that all 3 children will have blue eyes is $1/2 \times 1/2 \times 1/2 = 1/8$.)

148. No, the odds are stacked in Bonnie's favor. (There are 9 out of 21 ways to get a total of 6 or less, and 12 out of 21 ways to get a total greater than 6.)

149. 1/4 (The chance of getting heads is always 1/2. To get two heads, the chances are $1/2 \times 1/2 = 1/4$.)

150. 23 (The table shows all the possible combinations. Subtracting the one repeated combination—EU and UE—leaves 23 unique combinations.)

	S	Q	U	A	R	E
C	CS	CQ	CU	CA	CR	CE
U	US	UQ	UU	UA	UR	UE
B	BS	BQ	BU	BA	BR	BE
E	ES	EQ	EU	EA	ER	EE

151. 899 (The first digit has 9 possibilities, the second has 10 possibilities, and the third has 10 possibilities, or $9 \times 10 \times 10 = 900$, $900 - 1 = 899$.)

152. 7 (3 pieces of flavor 1, 3 pieces of flavor 2, and 1 piece of flavor 3.)

153. 106 $((254 + x)/4 = 90)$

154. 45

155. 250 $(4 \times 4 \times 4 \times 4 = 256, 256 - 6 = 250)$

Fractions, Ratio, Decimals, and Percents

156. Phantom, rock star, dragon, mermaid, vampire, jack-in-the-box

157. 18 gallons per minute (In line 1, the rate is $\frac{160}{16}$ = 10 gallons per minute; in line 2, the rate is $\frac{160}{20}$ = 8 gallons per minute. 10 + 8 = 18)

158. 4 hours and 10 minutes (C1—2 hours ÷ 4 = 30 minutes; C2—2 hours ÷ 3 = 40 minutes; C3—2 hours ÷ 2 = 60 minutes; C4—120 minutes; 30 + 40 + 60 + 120 = 250 minutes = 4 hours 10 minutes)

159. 120 days (If 10 people can paint 60 houses in 120 days, then 10 people can paint 1 house in 2 days, 5 people can paint 1 house in 4 days or 30 houses in 120 days.)

160. Answers will vary. Here is one solution:

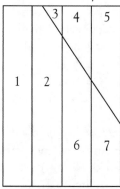

161. In theory, no, but in reality, yes. (Because the distance between them is small, and both Romeo and Juliet have mass, after Romeo has covered half the distance between them a few times, they will be touching, as their bodies will fill the remaining distance.)

162. One hundred-thousandth of a billion (one hundred thousandth of a billion—$10,000; one hundred million ten-millionths of a dollar—$10; one million millionths of a dollar—$1)

163. Division

164. 5 hours and 25 minutes (The amount of time it takes to fly from New York to Los Angeles has not changed.)

165. $\frac{3}{7}$ $\left(\frac{3}{2} \div \frac{3}{2} \div \frac{4}{3} \div \frac{5}{4} \div \frac{6}{5} \div \frac{7}{6} = \frac{3}{2} \times \frac{2}{3} \times \frac{3}{4} \times \frac{4}{5} \times \frac{5}{6} \times \frac{6}{7} = \frac{3}{7}\right)$

166. 6 (You can see a pattern start to form for the last digit in the decimal: 2, 4, 8, 6, 2, 4 . . .)

power	1	2	3	4	5	6
standard form	0.2	0.04	0.008	0.0016	0.00032	0.000064

167. $\frac{8}{48} + \frac{2}{6}$; $\frac{8}{24} + \frac{1}{6}$

168. No, the Big Bounce pogo sticks cost $9, while the Boing-Boing pogo sticks cost $12. (Big Bounce—$30 – ($30 × 0.70) = $9; Boing-Boing—($30 × 0.50) – ($15 × .20) = $12)

169. January 10 (70 beats per minute × 60 minutes in an hour = 4,200 beats in an hour, 4,200 × 24 hours = 100,800 beats in a day, 1,000,000 ÷ 100,800 ≈ 10 days.)

170. Cassie is 34 years old. (Because the age ratio was 1 to 35, Mrs. Pemberton is 34 years older than Cassie. This ratio has changed to 1 : 2, so, doubling the age difference will give 68. Cassie is half that, which is 34.)

171. 8 to 1

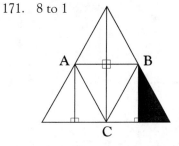

Daily Warm-Ups: Math Brain Teasers

172. $50 million ($1 - \frac{1}{2} + \frac{1}{4} + \frac{1}{5} = \frac{1}{20}$; the dog's inheritance is $\frac{1}{20}$ of the total. Since $\frac{1}{20}$ equals $5 million, the entire amount was $5 million $\times$ 20 = $100 million. The wife inherited $\frac{1}{2}$, which was $50 million.)

173. No (Lester counted each tank of fleas as a separate trial. In fact, he conducted only one trial. Rather than looking at each tank separately, he should average the results in all 5 tanks: (10 + 70 + 70 + 70 + 20)/5 = 48%)

174. π (The circumference of the smaller wheel is $C = \pi \left(\frac{7}{22} d \right) = \frac{22}{7} \times \frac{7}{22} \times d = d$. The circumference of the larger wheel is $C = \pi d$. So, the larger wheel circumference is larger by a factor of π.)

175. $\frac{11}{21}$ (The smallest gear's circumferences $C = \frac{1}{2} \pi d = \frac{1}{2} \times \frac{22}{7} d = \frac{11}{7} \times d$. The middle gear's circumference is $C = \frac{2}{3} \pi d = \frac{2}{3} \times \frac{22}{7} \times d = \frac{44}{21}$. The difference between the middle gear and the smallest gear is $\frac{44}{21} - \frac{11}{7} = \frac{11}{21}$.)

176. $\frac{1}{3}$ more times (Relate the rotation to the smallest gear. The largest gear will go around $\frac{1}{3}$ for every complete rotation of the smallest, and the second largest will rotate $\frac{2}{3}$ time for every complete rotation of the smallest. So, $\frac{2}{3} - \frac{1}{3} = \frac{1}{3}$. The second largest gear will rotate $\frac{1}{3}$ more than the largest gear.)

177. pentagon

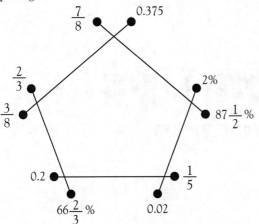

$\frac{7}{8}$ 0.375

$\frac{2}{3}$

$\frac{3}{8}$

2%

$87\frac{1}{2}$%

0.2

$\frac{1}{5}$

$66\frac{2}{3}$%   0.02

178. $\frac{5}{11}$ (The sequence is equivalent to $\frac{1}{11}, \frac{2}{11}, \frac{3}{11}, \frac{4}{11}$ alternating percent, fraction, decimal, so the next number should be $\frac{5}{11}$, written as a fraction.)

179. $0.\overline{7}$, $66\frac{2}{3}$%, 5/9, $0.\overline{4}$, $33\frac{1}{3}$% (These numbers are equivalent to ninths; that is, $0.\overline{7} = \frac{7}{9}$, $66\frac{2}{3}$% $= \frac{6}{9}$, $0.\overline{4} = \frac{4}{9}$, and $33\frac{1}{3}$% $= \frac{3}{9}$)

180. $\frac{1}{2}, \frac{1}{3}, \frac{1}{6}$ (Changing all the fractions to fractions with common denominators gives $\frac{30}{60}$, $\frac{20}{60}, \frac{15}{60}, \frac{12}{60}$, and $\frac{10}{60}$. The three fractions need to total $\frac{60}{60}$.)

Turn downtime into learning time!

For information on other titles in the

Daily _Warm-Ups_ series,

visit our web site: walch.com